PRINCIPLES

OF

DYNAMICS.

ELEMENTS.	FUNCTIONS.
Force = F.	Power P = F V.
Velocity = V.	Space S = V T.
Time = T.	Work K = F V T.

F : M = V : T.

MOMENTUM.

F T = M V.

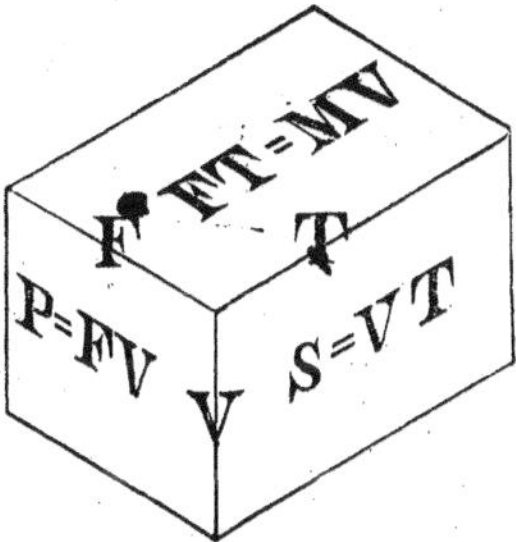

F : M = V^2 : V T.

WORK.

M V^2 = F V T.

THE LONGER WE TOIL,
THE MORE WORK WILL BE DONE,
BUT IF WE HAVE NOT TIME TO DO THE WORK,
IT WILL REMAIN UNDONE.

JOHN W. NYSTROM, C. E.

PHILADELPHIA:
JOHN P. MURPHY, PRINTER, N. E. FIFTH AND WALNUT.
1874.

To the Secretary of *the University of Michigan, Ann Arbor.*

Sir,

Accompanying this, you will be pleased to accept for the Library of

a Pamphlet just published, whose object is the establishment of "precision in the meaning of Dynamical Terms."

As all Learned Bodies must unite in a desire to realize this result, the author earnestly solicits a written opinion of your Institution, upon his treatment of the subject.

Awaiting your reply, he has the honor to be

Your Obedient Servant

John W. Nystrom

1010 Spruce Street.

Philadelphia, 1874.

PRINCIPLES

OF

DYNAMICS.

ELEMENTS.	FUNCTIONS.
Force = F.	Power P = F V.
Velocity = V.	Space S = V T.
Time = T.	Work K = F V T.

F : M = V : T.

MOMENTUM.

F T = M V.

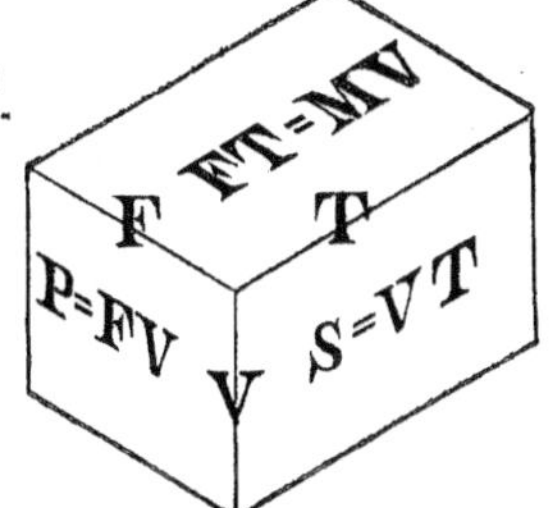

F : M = V²: V T.

WORK.

M V² = F V T.

THE LONGER WE TOIL,
THE MORE WORK WILL BE DONE,
BUT IF WE HAVE NOT TIME TO DO THE WORK,
IT WILL REMAIN UNDONE.

JOHN W. NYSTROM, C. E.

PHILADELPHIA:
JOHN P. MURPHY, PRINTER, N. E. FIFTH AND WALNUT.
1874.

ON

Establishing Precision

TO THE MEANING OF

Dynamical Terms,

AND

An exposition of the confusion which now besets the subject of Dynamics,

BY

JOHN W. NYSTROM, C. E.

PHILADELPHIA:
John P. Murphy, Printer, N. E. Fifth and Walnut.
1874.

To the Scientific World.

The earnest and indefatigable efforts of the writer to accomplish the purpose for which this Pamphlet is written, have exposed him to such frequent and long continued embarrassments, that in the interests of science, he sees but one expedient by which he can hereafter avoid them. That expedient is, to present his reasonings in printed form, to the various learned societies, in different countries, and to respectfully invite their judgments upon them.

The inexorable demands of Science, for precision in its vocabulary and perspicuity in its principles, are none the less felt and needed within the domain of Dynamics; and the clash of doctrine and dogma, as well as of definition, which have heretofore prevailed there, ought, if practicable, to be brought to an end.

If the present brochure can achieve so desirable a consummation, the writer will be abundantly gratified; whilst at the same time, he will accept with grateful acknowledgment, the arbitraments of scholars, even if they prove adverse. Should the subject, upon the basis here submitted, be thus definitively agitated and adjudged, it must necessarily result either in the maintenance or overthrow of the writer's views, and, in either event, rid the subject of those jarring authorities which have so long remained the opprobrium Professorium.

JOHN W. NYSTROM,

1010 Spruce Street,
Philadelphia.

PROCEEDINGS

OF THE

COMMITTEE ON DYNAMICS,

HELD AT

THE FRANKLIN INSTITUTE.

Committee.

JOHN W. NYSTROM, Chairman,
FAIRMAN ROGERS, Professor,
J. VAUGHAN MERRICK, Esquire,
LEONARD G. FRANCK, Professor,
† GEORGE F. BARKER, Professor,
JOHN H. TOWNE, Esquire.

First Meeting, on Saturday, November 8, 1873.

Members present—Prof. Fairman Rogers, J. Vaughan Merrick, John H. Towne, and the Chairman.

The Chairman called the meeting to order, and introduced the subject as follows:

GENTLEMEN,

This Committee was appointed at the September meeting of the Institute, for the purpose of establishing precision to the meaning of dynamical terms, and of selecting and adopting such terms as may be found proper, and of rejecting those that may be considered otherwise.

In order to utilize our time to the best advantage, to alleviate our labor, to avoid unnecessary repetitions of arguments, and enable us to arrive at decisive conclusions as far as we go, it may be advisable that the Committee should discuss the subject in the following order, namely;

Order of Discussion.

1st. To discuss and decide whether or not there exist between dynamical quantities, such distinction as elements and functions; and, if so, to classify them as such.

2nd. Assuming that dynamical quantities are classified into elements and functions, the Committee will then decide what elements the functions are composed of.

3rd. To discuss and decide upon the proper terms which will clearly distinguish dynamical quantities from one another, and give to each of them a perspicuous definition.

4th To discuss how far these adopted terms will cover all cases in dynamics; and to select and adopt such auxiliary terms as may be necessary, with perspicuous definitions.

5th. To reject all terms which may be found unnecessary, and to give reasons for so doing.

These five points of discussion will probably cover the principal object for which this Committee is appointed.

What action will the Committee take upon this proposed order of discussion?

The proposed order of discussion was approved, and the Committee requested that the whole programme should be read before commencing the discussion, whereupon the reading of the paper was continued.

The first point of discussion is to decide whether or not there exist between dynamical quantities, such distinction as elements and functions?

In chemistry, matter is classified into elements and binary compounds; but there exists no such distinction between dynamical or physical quantities.

In order to settle this question, it is necessary first to decide what is meant by a physical element, and by a physical function?

Definitions of the principle terms in dynamics are prepared for the Committee to act upon, and which are intended as materials for the report, as far as approved and modified in our proceedings.

The first eight definitions involves the fundamental principle of dynamics.

Principles of Dynamics.

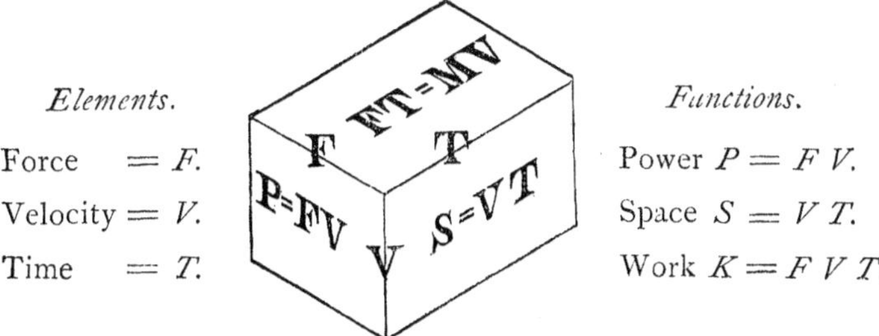

Definitions of Dynamical Terms.

Element, is an essential principle which cannot be resolved into two or more different principles.

Function, is the product of two or more elements.

A function is resolved, by dividing it with one or more of its elements.

Force, is any action which can be expressed simply by weight, without regard to motion or time; it is an essential principle which cannot be resolved into two or more principles, and is therefore an element.

Velocity, is speed or rate of motion; it is an essential principle which cannot be resolved into two or more principles, and is therefore a simple element.

Time, is duration, or that measured by a clock; it is an essential principle which cannot be resolved into two or more principles, and is therefore a simple element.

Power, is the product of the two first elements *force* and *velocity*, and is therefore a function.

Space, is the product of the two elements *velocity* and *time*, and is therefore a function.

Work, is the product of the three simple elements *force*, *velocity* and *time*; and is therefore a function.

Work, is also the product of the element *force*, and the function *space*, because the function *space* contains the elements *velocity* and *time*.

Work, is also the product of the function *power* and the element *time*, because the function *power* contains the elements *force* and *velocity*.

Corollary.

Force, in dynamics, means motive force, which invariably produces motion, power and work.

Velocity, is obtained by resolving the function space and eliminating the element time.

Motion, is continuous change of position in regard to assumed fixed objects, and is analogous to space.

We say that a body has more or less motion when it moves a greater or less space.

Space, in dynamics, means linear space, or more correctly, the *generation* of that space by velocity and time.

The square or cube of an element, is a function.

Horse-Power.

The unit *horse-power* was established by James Watt, to be equivalent to the power required in raising a weight of 33,000 pounds with a velocity of one foot per minute, or 550 pounds, one foot per second.

Man-power. The unit Man-power was established by Morin, to be 50 foot-pounds of power, which, compared with Watt's unit, will be 11 man-power per horse-power.

The capability of a horse is not a definite quantity, and the unit 33,000 pounds raised one foot per minute, is higher than the average performance of a horse, and the number 33,000 is unnecessarily difficult to handle in calculations and mental estimates of horse-power; for which reason it is recommended that the standard unit *horse-power* be established in the United States, as equivalent to 30,000 pounds raised with a velocity of one foot per minute, or 500 pounds, one foot per second; in which case there would be the even number 10 man-power per horse-power.

This unit would approach nearer to the average performance of a horse, and the numbers are more easily handled in calculation.

A velocity of only one foot per minute, as assumed in Watt's unit of horse-power, is too slow for a clear conception, while one foot per second is an appreciable quantity, for which reason it is more natural to use the expression *feet per second* in the unit of horse-power. The standard unit of power is foot-pound per second, of which there would be 500 per American standard horse-power.

This expression of horse-power, is used on the Continent of Europe, in which the Prussian horse-power is 513 foot-pounds.

$$\text{Watt's horse-power } HP = \frac{F\,V}{550}$$

$$\text{American horse-power } HP = \frac{F\,V}{500}$$

The difference between the two units of horse-power would be 10 per cent.

The average weight of a horse is ten times that of a man.

A horse eats ten times as much as a man.

The average power of a horse, is ten times that of a man.

Foreign terms and units for horse-power.

COUNTRIES	TERMS	ENG. TRANSLA'N	UNITS	ENG. EQUIV.
English	Horse-power	Horse-power	550 foot-pounds	550 foot-pounds
French	Force de cheval	Force-horse	75 Killogr. metres	542.47 foot pounds
German	Pferde-krafte	Horse-force	513 Fuss-funde	582.38 foot-pounds
Swedish	Häst-kraft	Horse-force	600 Skålpund fot.	542.06 foot-pounds
Russian	Sil-lochad	Force-horse	550 Fyt-funt.	550 foot-pounds

The quantities **force** and **power**, are clearly distinguished by different terms, only in the English language.

On the continent of Europe, power is called force, and the only distinction between the two quantities is by expressing power in foot-pounds, which also means work.

The word *force*, is needed in the Swedish and German languages.

In the French, Italian, Spanish, and Russian languages, there are words which correspond with *power*, but they are not used in that sense in dynamics, where the term *force* is used for *power*.

The Swedish and German word *kraft*, ought to be used for *power* only, and not for *force* as it is also used.

The words expressing *work*, are clear and definite in all languages.

Dynamics of Matter.

Matter is that of which bodies are composed, and occupies cubical space.

Mass is a group of matter, or the real quantity of matter in a definite body.

The *mass* of a definite body, is a constant quantity, and an essen tial principle which cannot be resolved into two or more principles, for which reason it is a simple element in dynamics.

We have, then, in physics, four fundamental and distinct simple elements, namely; **Force, Velocity, Time** and **Mass.**

The relation between these four elements is well known as an established fact, namely; **force** is to **mass** as **velocity** is to **time.** This is the fundamental analogy in dynamics of matter.

Denoting $M =$ mass, the algebraic expression will be

$$F : M = V : T.$$

Momentum of motion $M\,V = F\,T$, momentum of time.

Momentum of a moving body is thus the product of mass and velocity, which is a function, and which, accordingly, should not be confounded with the element force.

The momentum of a moving body is equal to the product of the two elements *force* and *time*, or in other words, the function which sets a body in motion, or brings a moving body to rest.

The force of a moving body is equal to the momentum $M\,V$, divided by the time T, in which the body is brought to rest.

$$F = \frac{M\,V}{T}$$

Weight is the force of attraction between the earth and the body weighed.

The force of attraction varies inversely as the square of the distance between the attracting bodies, wherefore, weight is not a constant quantity, but is proportionate to mass when compared in the same locality.

Weight and mass are, therefore, two distinct principles.

English and American units of measure for Dynamical Quantities.

Elements.

$F =$ force in pounds avoirdupois.
$V =$ velocity in feet per second.
$T =$ time of action in seconds.
$M =$ mass expressed in units of 32.17 pounds avoirdupois each.
$W =$ weight or force of attraction in pounds avoirdupois at the earth's surface.

Functions.

$P = F\,V$, expressed in foot-pounds of power.
$S = V\,T$, space, expressed in linear feet.
$K = F\,V\,T$, expressed in foot-pounds of work.
$M\,V = F\,T$, dynamic momentum, expressed in foot-pounds.

There are three kinds of foot-pounds in dynamics. When the expression *foot-pounds* is used, it should be stated whether it is foot-pounds of power, momentnm or work.

There is yet no definite unit established for measuring mass, for which reason it is recommended that the following unit and term be adopted.

From the fundamental analogy $F : M = V : T$,

we have the mass $M = \dfrac{F\,T}{V}$.

In the case of a falling body, the force of gravity $F = W$, the weight of the body.

Then the mass $M = \dfrac{W\,T}{V}$.

It is known from experiments that the velocity attained by a body falling freely under the action of gravity, for a time of one second, at or near the surface of the earth, is about $V = 32.17$ feet per second, which is generally denoted by the letter g.

Assuming $T = 1$, and $g = 32.17$, the units of mass can be expressed by

$$\frac{W}{g} = \frac{W}{32.17}.$$

Then a mass weighing 32.17 pounds at the surface of the earth, would be one unit of mass, which unit ought to be termed **matt,** from the word matter.

Work in moving bodies.

The work stored in a moving body, is that expended in bestowing its motion, and which is equal to the work restored by bringing the body to rest.

It has been decided that work $K = F\,V\,T$, in which V means the mean velocity of the force F in the time T.

In the case of a force setting a body in motion, its mean velocity is only one-half of the final or uniform velocity after the force ceases to act. Therefore, when V means the uniform velocity of a body, the work stored in it is only

$$K = \frac{F\,V\,T}{2}.$$

From the fundamental analogy $F : M = V : T$, we have the time

$$T = \frac{M\,V}{F}.$$

Insert this value of T in the formula for work, and we have,

$$K = \frac{F\,V}{2} \cdot \frac{M\,V}{F} = \frac{M\,V^2}{2}.$$

That is to say; the work stored in a moving body, is equal to half the product of the mass M and square of the velocity or V^2, which is a function, and accordingly cannot be expressed by the term force which is an element.

The above rule holds equally good for revolving bodies, only that the velocity V must be measured in the circle of gyration.

When the mass M is expressed by weight W, we have the work

$$K = \frac{W\,V^2}{2\,g} = W\,S = F\,S.$$

S = space or height which the weight is lifted or falls.

Force of Falling Bodies.

In the case of a body falling freely under the action of gravity, the work of the fall is equal to the work of the blow which brings the body to rest.

F = force of the blow, and s = space of penetration of the blow.

$$W : F = s : S, \text{ and } W\,S = F\,s.$$

Force of a falling body $F = \dfrac{W\,S}{s}$.

That is to say, the force of the blow of a falling body, is equal to the product of the weight W and fall S, divided by the space s of penetration.

Definitions of Auxiliary Terms.

Gyration, is a term applied to circular motion in dynamics of matter.

Centre of Gyration in a revolving body, is a point in which if all the moving matter were there contained, the body would present the same inertia as when the matter is distributed around that point.

Radius of Gyration, is the distance from the centre of rotation to the centre of gyration, which will hereafter be denoted by the letter X.

Circle of Gyration, is that described by the centre of gyration.

Inertia, is the incapability of matter or of a body to change its own state of motion or rest.

Force of Inertia, is equal to any force applied to change the state of motion or rest of a body, and is generally denoted by the letter I.

Moment of Inertia, is either mass, weight, volume or surface, multiplied by the square of its radius of gyration.

The term, *moment of inertia* is used in finding the radius of gyration, and is denoted by the letter E.

The sum of all the moments of inertia of all particles in a body, is equal to the moment of inertia of that body; or the sum of al the moments of inertia of all bodies in a system, is equal to the moment of inertia of that system.

Let a, b, c and d represent either masses, weights, volumes or surfaces in a rigid system rotating around an axis; and v, x, y and z, the respective radii of gyration. $Q = a + b + c + d$, and $X =$ radius of gyration of the system.

Then, the moment of inertia is

$$E = Q X^2 = a v^2 + b x^2 + c y^2 + d z^2,$$

of which the radius of gyration will be

$$X = \sqrt{\frac{a v^2 + b x^2 + c y^2 + d z^2}{Q}}.$$

This is the use of moment of inertia, which is a constant quantity in a revolving body, but is of no use after the radius of gyration is known.

Work in a Revolving Body.

It is decided that the work stored in a moving body, is

$$K = \frac{M V^2}{2}.$$

For circular motion, $V = \frac{2 \pi X n}{60}$, in which $n =$ revolutions per minute.

$V^2 = \frac{4 \pi^2 X^2 n^2}{60^2}$ and the work $K = \frac{M}{2} \cdot \frac{4 \pi^2 X^2 n^2}{60^2}$.

Work in a revolving body, $K = \frac{M X^2 n^2}{180.377}$, without regard to moment of inertia.

When the mass of a revolving body is expressed by weight W, the work will be

$$K = \frac{W X^2 n^2}{180.377 g} = \frac{W X^2 n^2}{5867.16}.$$

Bartlett's definition of moment of inertia, page 186; is as follows: "*Moment of inertia of any body, is twice the quantity of work exerted by its inertia, during a change in the square of its angular velocity equal to unity.*"

This definition is not intelligible. When it is required to know the work stored or restored during a change of angular velocity from N to n revolutions per minute of a rotating body, we have simply

$$\text{Work } K = \frac{W X^2}{5867.16}(N^2 - n^2),$$

without regard to moment of inertia.

Adopted Terms.

It is recommended that the dynamical terms herein defined, shall be adopted as standard terms.

It is also recommended that the letters herein used to represent terms and the corresponding quantities, shall be adopted as standard notations in all writings on dynamics.

Difficulties are frequently experienced in the use of dynamical formulas, for the want of a systematic notation, and very often, the true meanings of letters are not properly explained; whilst, with a standard system of notation, the formula is obvious at a glance.

First Meeting.

After the programme was read by the Chairman and discussed by the Committee, it was decided to send the paper to each member for revision, and that whatever remarks were made, should be written on the blank pages opposite the matter criticised.

The members of the Committee will send the paper amongst themselves in the following order:

The Chairman to Professor Leo. G. Franck, to Professor George F. Barker, to Professor Fairman Rogers, to Mr. J. Vaughan Merrick, to Mr. John H. Towne, to the Chairman.

The Meeting then adjourned.

---

202 W. RITTENHOUSE SQUARE.
December 5, 1873.

MY DEAR SIR:

As I fear that an important piece of business may prevent me from meeting your Committee to-morrow morning, I write to you my views on the subject in order that I may be represented.

After carefully considering the subject of the "Definitions of Dynamical Terms," I feel that the matter is so important and so surrounded by difficulties, that only a very long and careful discussion by a large and authoritative Committee, combined perhaps with extensive correspondence with scientific societies and scientific authors, could bring out a system which would be really valuable.

I do not think the present Committee has either the time or the disposition to undertake so great a labor, nor do I think that the Franklin Institute is exactly the body from which such a scheme should emanate.

These being my views I am not willing to accept the report prepared, as the report of the Committee, especially as I cannot agree to some of the points made in that report.

While believing that a certain confusion does exist in the terms and definitions in ordinary use, I think that the corrections can only come from a gradual improvement in the treatises on Mechanics, and could not be affected by the report of our Committee.

If I possibly can do so, I will try to be at the Institute before the Committee gets through its meeting.

Yours Respectfully,

FAIRMAN ROGERS.

MR. J. W. NYSTROM.

Rejected Terms.

It is proposed to reject the following terms in dynamics, on the ground that they are superfluous and confusing.

Vis-viva, literally translated, means *living force;* the term is used to denote double the work in a moving body, but there exists no such quantity. The term denotes what does not exist. Force of any kind is an element, and should not denote work, which is a function. The term *Vis-viva* conveys an idea which has often been entertained, namely, that a dead body can possess a virtue of life, which erroneous notion has caused much discordance in the elucidation of dynamics.

Energy. This term is used to denote work, but the sense of it conveys an idea of a different virtue, namely, that of activity or vigor, which is power. We say that a man has a great deal of energy, when he can accomplish much in a short time, which is a virtue of power.

The term *work* is the proper name for that function which is often erroneously termed energy.

Quantity of motion, is a term also used to denote work, which latter is a different function from that of motion. The sense of this term is inseparably associated with an idea of more or less space. (See Corollary.)

The terms **Actual, Total, Quantity, Mode, Potential, Intrinsic,** &c., are often appended to terms without affecting the nature of the quantity so denoted; the objection to which is, that one and the same quantity is differently defined according to the combination of appended terms. For instance, in Geometry we have the quantity volume, which corresponds to work in dynamics, and which may be expressed thus:

Volume of a cube.
Cubical volume of a sphere.
Total intrinsic volume of a cone.
Actual potential volume of a cylinder.
Actual total quantity of volume of a pyramid, &c.

It is all simply **volume,** like the different combinations of terms denoting work, means simply **work,** or the product of the three simple elements, *force*, *velocity* and *time*.

The following terms which are proposed to be rejected, are taken from the works of **Rankine, Bartlett** and **Moseley.**

Rejected Terms.	Reason for Rejection.
Acting force.	All forces act.
Force of motion.	Means motive force.
Working force.	" " "
Quantity of moving force.	" " "
Quantity of motion.	Has no definite meaning.
Mode of motion.	" " " "
Mode of force.	" " " "
Moment of activity.	" " " "
Mechanical power.	Means simply power.
Mechanical effect.	" " "
Quantity of action.	" " "
Efficiency.	" " "
Rate of work.	" " "
Dynamic effect.	Used for power or work.
Quantity of work.	Means simply work.
Total quantity of work.	" " "
Actual total quantity of work.	" " "
Total amount of work.	" " "
Actuated work.	" " "
Vis-viva.	" " "
Living force.	" " "
Energy.	" " "
Actual energy.	" " "
Potential energy.	" " "
Energy of motion.	" " "
Energy of force.	" " "
Heat a form of energy.	" " "
Heat a mode of motion.	" " "
Mechanical potential energy.	" " "
Quantity of energy.	" " "
Stored energy.	" " "
Intrinsic energy.	" " "
Total actual energy.	" " "
Work of energy.	" " "
Equation of energy.	Formula for work.
Equality of energy.	Primitive and realized work.
Transformation of energy.	" " " "

Second Meeting.

The Committee on dynamics held its second meeting at the Hall of the Institute on the 6th of December 1873.

Members present. Professor Fairman Rogers, J. Vaughan Merrick, Professor Leonard G. Franck, John H. Towne and the Chairman.

Since the last meeting, each member of the Committee has examined the programme of discussion, and made several remarks on the definitions of dynamical terms.

The substance of the remarks are, that "*velocity* is not an element," that "*space* is an element," and that "*work* is composed of two elements, namely *force* and *space*." *Time* is thus left out as an element of work. It was also remarked that "*time* is an element of *power*.

The Committee decided not to discuss the subject any further, but to refer the same to the Smithsonian Institution; whereupon Mr. J. Vaughan Merrick made the following motion, which was seconded by Mr. John H. Towne, and carried:

Resolved, That the report made by the Committee shall embody the following material,—

1st. The fact that great indistinctness exists among scientific Bodies and writers here and elsewhere in the use of scientific terms in dynamics—and a list of such terms by way of explanation.

2nd. That the Franklin Institute being a body largely composed of persons engaged in the mechanic arts, as also in instructing students in mechanics, experience great difficulty in carrying out its objects with efficiency, from the indistinctness referred to. Hence it is believed,

3rd. That it is highly desirable to determine the meaning of such terms, and to eliminate all which may be found unnecessary.

4th. That such a work can best be accomplished by a national body of authority, known throughout the scientific world.

5th. That the Smithsonian Institute being such a body, this Committee recommend that the Franklin Institute request it to consider the subject, and take such action as it may deem proper to bring about this desirable object.

The resolution was adopted and the meeting adjourned.

Third Meeting.

The Committee was called to meet at the Institute on Saturday, December 13, 1873, but only Professor Fairman Rogers and the Chairman came to attend.

A report had been prepared in accordance with the resolution adopted at the last meeting, which was expected to be signed and submitted to the Institute at the December meeting, but was not accomplished.

† Professor George F. Barker at first consented to serve, but has not attended any of the meetings, and nothing has been heard from him in regard to the subject of the Committee.

Note 1.

The Chairman asked the Committee to recommend the publication in the Journal of the Franklin Institute, of the programme prepared for discussion, which was not consented to, on the ground that it might appear as if the Committee endorsed its ideas.

Note 2.

The writer publishes the proceedings of the Committee for distribution among Scientific Institutions and other Learned Societies at home and abroad, in the hope that his principles of Dynamics will be understood and approved somewhere.

Note 3.

A manuscript embracing a complete treatise on "*Elements of Mechanics, classified with precision as to the meaning of Dynamical Terms,*" and based upon principles laid down for the Committee on Dynamics of the Franklin Institute, was finished in the year 1864; but Learned Men who are considered as high authorities and judges on the subject, have, it appears, formed a formidable barricade against the publication of that manuscript.

The writer has demonstrated to the publishers, that these high authorities have not been able to sustain their objections in open contests on the subject; but all in vain.

Note 4.

The paper on Dynamics submitted to the National Academy of Sciences, in the year 1865, was substantially the same as the programme prepared for the Committee of the Franklin Institute.

The Academy declined to act upon that paper, on the ground that the subject was considered to be perfectly clear in the text-books already.

Note 5.

Discussions on the Principles of Dynamics, in the Journal of the Franklin Institute.

AUTHORS	YEARS	MONTH	PAGE
John W. Nystrom,	1864	Nov.	325
Meeting of the Institute,	1864	Nov.	356
Prof. De Volson Wood,	1865	Jan.	27
John W. Nystrom,	1865	March	181
Prof. De Volson Wood,	1865	June	385
Dynamometer, Stockholm,	1865	June	392
John W. Nystrom,	1865	July	56
Prof. De Volson Wood,	1865	Sept.	177

Note 6.

The principles of Dynamics were discussed by the Engineers in the Bureau of Steam Engineering, Navy Department, Washington, D. C., in the year 1865.

In favor of Nystrom.	*Opposed to Nystrom.*
Stephen D. Hibbert,	Benj. F. Isherwood,
Albert S. Green,	George M. Green,
Albert Aston,	And several others.
F. C. H. Ramsden.	

Note 7.

A complete set of formulas based upon the herein described principles of dynamics, is published in Nystrom's Pocket-Book of Mechanics, twelfth edition.

ON

FORCE OF FALLING BODIES

AND

DYNAMICS OF MATTER,

CLASSIFIED WITH PRECISION TO THE MEANING OF DYNAMICAL TERMS.

BY

JOHN W. NYSTROM, C.E.

PHILADELPHIA:
J. B. LIPPINCOTT & CO.
1873.

Entered according to Act of Congress, in the year 1872, by
JOHN W. NYSTROM,
In the Office of the Librarian of Congress at Washington.

LIPPINCOTT'S PRESS,
PHILADELPHIA.

FORCE OF FALLING BODIES.

A DISCUSSION on this subject has originated in the *Scientific American,* from a question published in that journal, June 8th, 1872, as follows:

" *Force of Falling Bodies.*—We have a steam-hammer weighing exactly three tons, including piston and rod; the stroke is four feet, and the hammer falls by its own gravity. What will be the force of the blow, making no allowance for friction? What is the formula for the calculation?

" (Signed) "J. E."

The following answers were given:

[*From the Scientific American, July* 6, 1872.]

Force of Falling Bodies.

To J. E., query 12, June 8.—The hammer will strike with a momentum of 160,164.5472 pounds. The formula is

$$\text{the square root of } (4 \times 64.33) = 16.0312 \text{ velocity.}$$

Then

$$4.426 \times 6000 \times 160312 = 160164.5472.$$

Or, multiply the fall in feet by 64.33; the square root of the sum is the velocity; and multiply the weight in pounds by 4.426 and that by the velocity, and you have the momentum.

E. E. W., of W. Va.

[*From the Scientific American, July* 13, 1872.]

Force of Falling Bodies.

If J. E., query 18, page 385, last volume, will multiply the weight of any falling body, in pounds, by the height of the fall in feet, he will have the force of the blow in foot pounds. Leaving friction out of the question, the force of the blow of his hammer is precisely equal to the force expended in raising it, namely, $6000 \times 4 = 24{,}000$ foot pounds. Converted into heat, this force would be competent to raise the temperature of one pound of water a little more than 31°, thus: 24,000 divided by 772 equals 31.09 units of heat.

W. H. P., of Iowa.

[*From the Scientific American, August* 10, 1872.]

Force of Falling Bodies.

In view of the difference between the two answers to J. E., query 12, June 8, and of my own ideas, somewhat different from either, I would say: The striking force of a moving body, in whatever direction it moves, is its momentum. Its momentum is the joint result of its quantity of matter and its velocity. The ratio of this momentum to that of other moving bodies is compounded of the ratio of its quantity of matter, which is indicated by its weight, and of its velocity at the instant in question. Its momentum, therefore, is not weight any more than it is space or time, and it cannot be expressed by pounds, in the ordinary sense of that word, any more than by feet or by seconds, nor is it expressed by any two of those terms. To obtain a statement of the momentum of a body for the purpose of comparison: Multiply its weight by its velocity—its number of pounds, for instance, by the number of feet it would move in a second if it should proceed for a second at the rate for the instant in question. The velocity of a falling body is continually accelerated, and it increases not as the space fallen over but as the square root (query? Ed.) of that space. Therefore, to multiply the weight by the space fallen over, will not give the momentum. The velocity of a falling body at the end of one second of its fall is $32\frac{1}{6}$ feet per second, and it has fallen one-half that distance. It will fall $4\frac{1}{48}$ feet in half a second, and its velocity is then $8\frac{1}{24}$ feet in half a second. The velocity at four feet descent is nearly the same, but more exactly is 16.0312 feet per second. This multiplied by the weight in pounds gives the momentum. The general formula is: The square root of (64.33 multiplied by the distance fallen) = the velocity, and the velocity multiplied by the weight = the momentum. So much for determining the momentum. The extent of change produced by the blow of a hammer has a compound relation to the force of the blow and the ability of that which it strikes to resist. Some obstacles resist in proportion not only to intrinsic power, but also to the time during which they exert their resistance, and their resistance to a blow is less as the velocity of the blow is greater. Such are the different attractive, repulsive, and expansive forces, and such is substantially the case where springs are to be bent and where many forms of cohesion are to be overcome. In such cases, the change produced is as the weight multiplied by the square of the velocity, and in case of a falling body is as the weight multiplied by the distance fallen. Other resistances are independent of time, and are in proportion to the space over which the resistance operates. Such is substantially the case of friction. Here the change is as the momentum of the blow. It is so in the case of bodies resisted by the momentum or inertia of other bodies, or, as in greater or less degree is the case of a body moving through liquids, of the particles of bodies. The case of forging with a hammer presents a compound of both these kinds of resistance, varying in their proportions with the nature of materials, degree of heat, and other considerations.

G. M. T.

[*From the Scientific American, August* 10, 1872.]

Force of Falling Bodies.

To the Editor of the Scientific American:

Since you are publishing a series of articles on "Weight, Pressure, Power, Force," etc., it would be useful to so explain the acting force of a body in motion, its momentum or striking force, that, if such a thing be possible, your readers may understand by what means, by what it is measured, and how determined.

While this is one of the simplest problems in physics, as well as one of the most essentially practical, it is one of those of which the majority of the people are most profoundly ignorant, as is shown by the frequent questions on the subject in your valuable paper, and by the replies, no two of which are alike, and which indicate that the correspondents are hopelessly befogged.

In your number of July 6, page 10, a correspondent—misled by Haswell, probably—estimates the force of the hammer, weighing three tons and falling four feet, at over 160,000 pounds. But what does he mean? What is a pound of force? To what is it equal? What work will it do? He does not say foot pounds, and if he means that, he is wide of the mark in his estimate. A blow cannot be compared with weight or pressure alone.

It should be universally known, if possible, that force is estimated by the measure of the work it is competent to perform, the number of pounds it will raise one foot high. The force which will lift one pound one foot is called a foot pound, and is the unit used to express the amount of a force. Gravitation, being a constant quantity, is a convenient standard, and force measured by the amount of gravitation it will overcome affords a statement quite intelligible to any intelligent person. Next, it should be known that this same one pound, in falling freely one foot, will accumulate the same amount of force, that is, gravity will impart to it in its descent the same amount of force which it took from it in its ascent, and therefore the force of the blow will be just one foot pound; and, if converted into heat, would produce exactly the amount of heat which would be required to lift the one pound one foot high again.

In general, the force with which any falling body will strike is precisely the amount required to lift the same body to the height from which it fell. When, therefore, the weight and height are given, their product is the force of the blow in foot pounds, and in the case of this hammer, would be $6{,}000 \times 4 = 24{,}000$ foot pounds. The force of a "weight of one pound falling two feet" would be $1 \times 2 = 2$ foot pounds, while Haswell's "Engineer's and Mechanic's Pocket Book," page 419, gives it at 11.34 pounds, whatever that may mean.

If the velocity is given, we find the height as follows: Dividing the velocity by $32\frac{1}{6}$ (the velocity acquired in each second) gives the time of fall in seconds, and multiplying the square of the time by $16\frac{1}{12}$, we have the height from which the body must have fallen to acquire the given velocity, which, of course, is also the height to which the body would ascend, if projected upward with the same initial velocity before its force would be expended in overcoming gravitation. Obviously, the force of the blow will be the same, with the same velocity, whether the motion be downward, upward, or horizontal; hence, to find the force with which it is moving, we only require to find the height from which a body must fall to acquire the given velocity, and said height, multiplied by the weight, gives the striking force in foot pounds, or the amount of work the body would perform, the resistance it would overcome, the weight it would lift one foot, or the heat it would produce; and also, what is the same thing, we have the amount of force expended in imparting to the body the given velocity.

The general confusion of ideas upon this subject is probably largely due to the fact that the text-books differ widely, and the majority of them are entirely wrong, as they almost all teach that the striking force is proportional to the velocity, whereas it is, in fact, proportional to the square of the velocity, as is readily shown by the law of falling bodies enunciated in the very same books.

The formula above given is far more simple than the various arbitrary and fantastic ones so often presented by your correspondents, and has the peculiarity of being correct, and consequently consistent with all the laws of motion; and if you will give me space for a few examples, I believe its application will be perfectly plain to your readers. Instead of dividing the velocity by 32.16 and multiplying the square of the quotient by 16.08, we may, of course, obtain the same result by the shorter process of dividing the velocity by 8.02, and squaring the quotient.

1. A one pound ball moves 1000 feet per second; $(1000 \div 8.02)^2 = 15{,}545$. Its force then is 15,545 foot pounds, and as it weighs one pound, if its motion were directly upward it would mount to the height of 15,545 feet, and on returning would acquire in its descent the same velocity of 1000 feet. The force expended, then, in imparting this velocity was equivalent to that required to raise 15,545 pounds one foot.

2. A twenty-four pound ball has a velocity of 50 feet per second; $(50 \div 8.02)^2 \times 24 = 931.44$ foot pounds. If this twenty-four pound weight were a hammer with a stroke of 38.81 feet, it would acquire a velocity of 50 feet, and would strike with a force of $38.81 \times 24 = 931.44$ foot pounds, and this amount of force, in any available form or mode of manifestation, would be sufficient to impart a velocity of 50 feet to a mass of 24 pounds, or to lift 24 pounds 38.81 feet, or to lift or throw one pound 931.44 feet high, or 931.44 pounds one foot high. In these calculations, there is no allowance made for atmospheric resistance.

W. H. PRATT.

DAVENPORT, IOWA.

Neither of those four communications answers the question of Mr. J. E., which is, *What will be the force of the blow?*

They state, how many foot-pounds, how much work, and how much momentum there is in the hammer, neither of which is force, as is required by the question. As I considered it difficult for the inquirer to make out anything from those answers, I volunteered to answer the question in my own way, as follows:

[*From the Scientific American, August* 31, 1872.]

Correspondence.

The Editors are not responsible for the opinions expressed by their Correspondents.

Force of Falling Bodies.

To the Editor of the Scientific American:

The question "With what force does a falling body strike?" has been frequently repeated in the *Scientific American* for the last twenty-five years, and has generally been answered by the batch of dynamical terms used in colleges and styled "scientific." The answers have invariably made the problem more obscure. Each one generally says that "the problem is very simple," and he pretends to understand the subject perfectly. I am one of those pretenders, and propose to answer the question in my own way, reference being made to the accompanying figure.

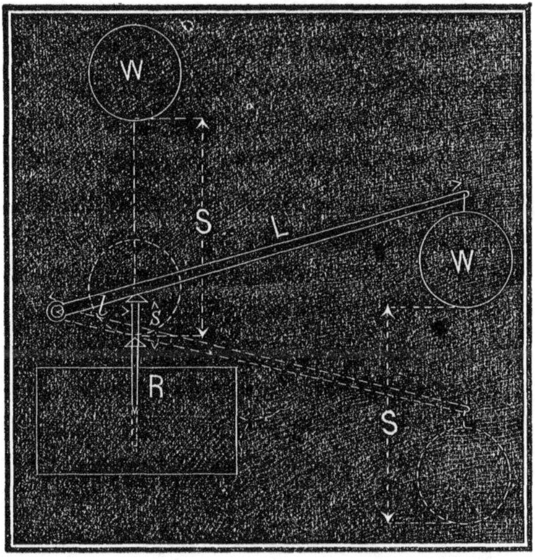

Let us assume the case of driving a nail into a piece of wood by the aid of a lever whose fulcrum is at *C*. The applied force is represented by the weight *W*, acting on the lever *L*. Let *R* denote the force of resistance in the wood, expressed by the same unit of weight as that of *W*, say pounds.

The weight *W*, acts on the long lever *L*, and the resistance *R*, on the short lever *l*. Then

$$R : W = L : l, \text{ and } R = \frac{WL}{l}$$

That is to say, the force of resistance in the wood is to the weight or force W, as the long lever L, is to the short lever l.

Let S represent the vertical height which the weight W moved, and s the distance which the nail was driven into the wood. Then

$$R : W = S : s, \text{ and } R = \frac{WS}{s}.$$

That is to say: the force of resistance in the wood is to the force or weight W, as the height S, is to the distance s.

Now let the same weight W, fall from an equal height S, directly upon the head of the nail, and the latter will be driven into the wood the same distance as by the aid of the lever. Therefore: the force with which the falling body acted upon the nail is to the weight of the falling body as the height of fall is to the distance the nail is driven into the wood. The force of the falling body is equal to its weight multiplied by its height of fall, and the product divided by the distance which the nail is driven into the wood.

JOHN W. NYSTROM.

PHILADELPHIA, PA.

My explanation on force of falling bodies was not well appreciated by an employee of the *Scientific American*, Dr. Vander Weyde, who made the following remarks on the subject:

[*From the Scientific American, September* 14, 1872.]

Correspondence.

The Editors are not responsible for the opinions expressed by their Correspondents.

Force of Falling Bodies.

TO THE EDITOR OF THE SCIENTIFIC AMERICAN:

I see, in an article on page 131 of your paper, that Mr. John W. Nystrom acknowledges himself to be "one of those pretenders" who think that they "understand perfectly the subject" of measuring the force of a falling body by taking, as unit of measurement, the mere weight of matter without motion. I desire here to say to him and to all those interested in the important question of measuring forces, in consideration that force must be distinguished from mere weight, that weight is merely a measure for an amount of matter, for a mere mass, and for nothing else; such weight, of course, is caused by gravitation, and thus can exert pressure, but as long as the weight does not produce motion, there is no force generated; therefore strictly speaking gravitation is no force, notwithstanding the conventional way of speaking of the force of gravitation; however, gravitation can beget force, and only does so in case it is allowed to produce motion. According to the modern conception of force, it is not something immaterial, independent of matter, but absolutely nothing but matter in motion. This motion may be hidden, molecular, when the force manifests itself as heat, electricity, etc., or the motion may be in the masses, when the force is directly measurable by two elements, the mass and the velocity. Accepting the customary symbols for these two different elements, the different degrees of force are expressed by the formulæ $v \times m$ and $v^2 \times m$, which are both correct according to circumstances. In the case of the effect of a blow produced by a falling body, for instance, the driving in of a nail, the identical case represented on page 131, the latter formula, corresponding with the theory of the *vis viva* (see any text-book on mechanics), must be applied. This is the first point in which the formulæ of Mr. Nystrom are faulty, as they are based on the lever, and thus not on the square of the velocity or space, but on the simple velocity : $v \times m$.

The result of this law of the *vis viva* is that, where gravitation increases or decreases, and with it the velocity of the falling body, the force of the blow will increase or decrease as the square of the gravitation, while the weight of the body will only increase or decrease in the simple ratio of the gravitation. Mr. Nystrom's figure and formulæ fail to take any account of this whatsoever.

But let us consider the expressions $v \times m$ and $v^2 \times m$ theoretically. It is evident that they have no value at all as soon as either of the quantities v or m becomes immeasurably small or disappears. Let, for instance, in the function $v \times m$ or $v^2 \times m$,

m become $=0$; then we have $v \times 0 = 0$ and $v^2 \times 0 = 0$, which conform to practical experience, because a blow with a mass equivalent to nothing must necessarily amount to nothing. Let, inversely, v be $=0$, and we have

$$0 \times m = 0 \text{ and } 0^2 \times m = 0,$$

again equivalent to nothing; a mathematical proof that a mere mass without velocity (motion) cannot possibly be reckoned equivalent to any force; and we see here the great mistake, thus far made by the authors of many text-books, in speaking of a force of, say, 100 pounds, or a ton.

The cause of this error is mainly to be found in the fact that a mere weight by its pressure will in some cases produce results similar to that of a force or blow. If, however, we attempt to measure force (matter in motion) by mere weight (matter in rest), we must continually fail and obtain incongruous results, as they are two incomparable quantities. This confounding of an actual force produced by a moving mass with mere weight or pressure produced by a stationary body, is the cause of fifty per cent. of the attempts, continually being made by the half educated, to find perpetual motion.

Now for a few practical illustrations: With a comparatively light hammer, we may easily drive a nail into a brick wall; if we try to do it by mere pressure, we shall crush the nail, or, to take Mr. Nystrom's own illustration, we can drive a nail into a board by the blow produced by dropping the head of a hammer on it from a suitable height, directed by guiding pieces, as in a pile driver; but take a similar nail, place it on the same board, attach the lever of proper length and hang the hammer head at the end of the lever, following practically the figure on page 131, and see if the nail will penetrate at all. If Mr. Nystrom had tried the experiment, he surely would never have taken the trouble to illustrate and publish his explanation.

The blow or percussion gives to a mass a shock, transmitted through it with the same velocity as a wave of sound would travel in that same mass; when the blow is violent and there is somewhere a want of continuity, or lack of strength, which prevents the wave from pursuing its course, its power will be expended there in crushing the material. This is the case in driving a nail. The motion will not be communicated to the board, but the force will be expended in crushing and cutting the fibres of the wood under the nail, so as to allow it to enter, while a weight or pressure placed on the nail will have plenty of time to communicate itself to the whole board. A striking illustration of this may be had when balancing a heavy board on its centre; it is then possible to drive a well-pointed nail with a smart blow deep in the board without moving the latter, while the same nail with a weight on top will scarcely make a mark on its surface, but will move the whole board. A pistol ball may be fired through a door without moving it on its hinges, which latter may be done by the slightest pressure of the finger. Scores of other familiar examples may be adduced, all proving the immense difference between force and mere pressure, and it is only to be wondered at and at the same time deplored that still so much confusion prevails in regard to this all-important subject.

P. H. VANDER WEYDE.

NEW YORK CITY.

The inquirer Mr. J. E., and other readers of the *Scientific American*, will no doubt be well informed on force of falling bodies, after having read Dr. Vander Weyde's philosophy on that subject. When I read the article I could not credit the possibility that Dr. Vander Weyde has ever graduated as doctor of philosophy in any creditable college; which, however, would make no difference to me if he has or not, as long as he keeps himself within his due limits. When an individual bears the title of Doctor or Professor we expect him to issue authority; but when he imposes upon the public with quack philosophy, as Dr. Vander Weyde has done, we have reason to suspect that there may be something wrong about his title, and we are justified in asking the doctor to show his diploma, and in holding him responsible for such offences. With this motive I addressed the editor of the *Scientific American* on the subject, and received the following answer:

[*From the Scientific American, September* 28, 1872.]

Who is Dr. Vander Weyde ?

During the past few weeks, an esteemed correspondent, J. W. Nystrom, Esq., C. E., of Philadelphia, has furnished to our readers several interesting communications, some of which have been answered and criticised by another of our valued correspondents, Dr. P. H. Vander Weyde, of this city. From the tenor of the following letter it would seem that our Philadelphia correspondent is a little suspicious of the respectability of his antagonist. But we can assure him that, in Dr. Vander Weyde, he has a foeman worthy of his lance.

To the Editor of The Scientific American:

Sir:—Will you be kind enough to inform me, through the *Scientific American*, if Mr. P. H. Vander Weyde, of New York, has a doctor's diploma, and if so, from which college he has received that title ? And what kind of a doctor is he ?

The answer to these questions will greatly oblige yours very respectfully,

John W. Nystrom.

1010 Spruce Street, Philadelphia, September 7, 1872.

We would inform our correspondent that Dr. Vander Weyde is a physician of the strictest orthodox sect; that he is an honored graduate of the New York University Medical College, of which John W. Draper, LL.D., is President; that he holds the regular diploma of that institution; that he enjoys the fellowship and esteem of many of our leading physicians and prominent men of science; that he is a native of Holland, where he received a university education; took the degree of Doctor of Philosophy in 1840; was the editor of a scientific periodical; in 1845, at Amsterdam, he received the honorary prize, consisting of the gold medal of the Society of Sciences, for his essays upon natural philosophy.

Dr. Vander Weyde is now a citizen of the United States. From 1859 to 1864, he was Professor of Physics, Higher Mathematics and Mechanics at Cooper Institute in this city. During nearly the same period, he was also Professor of Chemistry in the New York Medical College. From 1864 to 1866 he was

Professor of Industrial Science in Girard College, Philadelphia, Pa. His contributions to the scientific literature of the day have been very extensive, and are widely known.

These are only a few of the items of Dr. Vander Weyde's public record. But they are sufficient, we trust, to satisfy the inquiries of our correspondent, and remove from his mind any adverse prejudices that he may have formed concerning the qualifications of the distinguished gentleman whose public standing he has questioned.

Upon the authority of the *Scientific American* it appears that the said Vander Weyde has really graduated as doctor of philosophy in some college or university in Holland, which, however, I shall not feel convinced of until I have seen his diploma, or am informed of the fact direct from that university.

Holland stands very high in sciences, in fact, on the level with other nations, and its colleges are very strict in their studies and examinations, in which it is doubtful whether Dr. Vander Weyde's philosophy on dynamics of matter could have passed for good.

I cannot enter into any discussions in the *Scientific American* with such a philosopher as Dr. Vander Weyde has proved himself to be, for it is the duty of the university in Holland from which he bears his diploma (if he has any such) to have instructed him in the sciences in which he has proven himself deficient.

As soon as I can find out the name of that university, I shall endeavor to have the subject attended to, and if that university cannot endorse and sustain its doctor's philosophy, he ought to be called back and made to study over again, or be requested to return his doctor's diploma.

I have learned that he has been professor of natural philosophy in Girard College, Philadelphia; and if he has advanced such philosophy there as that he has done on force of falling bodies, his students are simply deceived.

If Dr. Vander Weyde had limited himself within the confusion which really exists in the want of precision to the meaning of dynamical terms, I should have had forbearance with him; but he dictates a doctrine which is contrary to established facts in physics.

It may be considered that I am rather severe upon Dr. V. W., but there are so many of this kind of philosophers that it will do no harm to tell the truth to one of them every now and then; and what has been said about Dr. V. W. is equally applicable to all the rest of the high authorities who have invariably attacked me in the style of quack der Weyde. When my ideas differ from what is written in their books, they blindly suppose that I am wrong, and they attack me with irrelevant philosophy, by which the public has been juggled, and I have been embarrassed all my life by quack opinions of high authorities. In many cases I have doubted whether the high authorities have themselves believed in their own statements, but they evidently expected me to have faith in their profound reasonings of perfect nonsense.

The object of this writing is, however, not to attack Dr. V. W., nor to prove that I am right and that he is wrong, all of which is of secondary importance to me and to the public; but my principal object is to call the serious attention of scientific institutions to the confusion in dynamics, and to the annoyance which that confusion causes to the public, and between individuals.

For the last ten years I have had repeated discussions on the confusion of dynamical terms, in which my opponents have not been able to sustain themselves. Some of these discussions are published in the *Journal of the Franklin Institute*, and also in the *Scientific American* for the year 1865. The high authorities generally maintain that work is independent of time!!! which is contrary to the opinion of manufacturers in regard to the eight or ten hours' struggle for a day's work.

We have no good text-books on dynamics, and, even in colleges, the subject is confused, not so much in substance as in terms, and the result is that graduated students, even from the same college, differ as to which is which in dynamics. The confusion is, however, acknowledged by high authorities, who have not been able to sustain themselves against me in open contests, and, when such authorities are professors in colleges, the public cannot thus receive proper information, and the subject remains obscure, as proven in the case before us. Under such circumstances I have a right to claim the issue of authority, which is as follows:

CLASSIFICATION OF DYNAMICAL TERMS.

I do not acknowledge the existence of more than one kind of force in physics, and that is, that action which can be expressed simply by weight, without regard to motion, time, power, or work. Force is derived from a great variety of sources; but when it is simply force, it can always be expressed by weight.

Force, motion, and time are simple physical elements. Space, power, and work are functions of those elements.

Elements.			*Functions.*	
Force $= F$	. . . .	1.	Space $S = VT$	4.
Motion $= V$	. . . .	2.	Power $P = FV$	5.
Time $= T$	. . . .	3.	Work $K = FVT$	6.

The weight of a body is the force of attraction between that body and the earth; and as the force of attraction varies inversely as the square of the distance between the centres of the attracting bodies, the weight of a defined body is not a constant quantity.

Mass means the real quantity of matter in a defined body, and it is a constant quantity which cannot be increased or diminished by force. Mass is generally denoted by the letter M, and is one of the four elements which constitute the dynamics of matter.

The relation between these four elements is well known and accepted as an established fact, namely,

$$M : F = T : V \quad . \; . \; . \; . \; . \; . \; . \; . \quad 7.$$

This is the fundamental analogy in the dynamics of matter, of which the two functions,

Momentum of motion, $MV = FT$, momentum of time, 8.

The function MV has been termed *force*, which is a great error, for it only denotes the product of the force and time consumed in giving the mass M the velocity V, and which is equal to the product of another force and time required in bringing the mass from motion to rest. There is no relation between the two forces, which are entirely governed by their respective times of action.

The function 6 expresses the

$$\text{Work } K = FVT \quad . \quad . \quad . \quad . \quad . \quad . \quad . \quad 6.$$

Multiply the momentums function 8, by the velocity V, and we have the

$$\text{Work } K = MV^2 = FVT \quad . \quad . \quad . \quad . \quad . \quad 9.$$

We see here that the function MV^2 means work, but it has been termed *force* and *vis viva*, which are also erroneous. MV^2 only denotes the product of force, motion, and time, which is the work consumed in giving the mass M the velocity V, and which is equal to the product of another force, motion, and time, or the work required or executed in bringing the moving mass to rest.

The function 4 expresses the

$$\text{Space } S = VT \quad . \quad . \quad . \quad . \quad . \quad . \quad . \quad . \quad . \quad 4,$$

which, inserted in function 9, will be the

$$\text{Work } K = MV^2 = FS \quad . \quad . \quad . \quad . \quad . \quad . \quad 10,$$

in which the function MV^2 denotes the product of the force and space in which the mass M attained the velocity V, and which is equal to the product of another force and space in which the moving body is brought to rest, and which is the same as the primitive function FVT.

None of these functions should be called force.

The present confusion in dynamics consists in that the functions are called force, for which we can make neither head nor tail of it.

Dynamics is now in the same condition as geometry would be if there was no distinction between *line*, *surface*, and *solid*, and it will remain in that confusion as long as the colleges teach that MV and MV^2 are forces.

The erroneous idea that MV and MV^2 are forces has stuck so tightly into the heads of philosophers as to cause an epidemic in dynamics.

Dr. Vander Weyde says: "Accepting the customary symbols for these two different elements, the different degrees of force are expressed by the formulæ $V \times M$ and $V^2 \times M$, which are both correct according to circumstances." I would ask Dr. V. W. if the circumstances depend upon the weather.

The *Scientific American* of the 22d of June, 1872, gives the following ideas of dynamics:

Weight, Pressure, Force, Power, Work.

The fact that the above words are often confounded together, for the simple reason that their true meaning is not well understood, has been the cause of many fruitless attempts at mechanical inventions and improvements. Most searchers for perpetual motion make no distinction between pressure and force, and are under the delusion that mere pressure can produce work, and we have seen writers on mechanics and we have even heard lecturers on scientific subjects speak of a force of, say, two tons weight. Weight alone is not force, neither is pressure equivalent to work; and it may therefore be useful to attempt some clear definitions of the above terms, in order to protect inventive minds against mistakes in mechanical reasoning.

Weight is simply the measure of an amount of matter referred to a certain standard accepted as a unit. This unit may be a gramme, a pound, a ton, or our whole earth, which the astronomers use; but, in either case, it conveys to the mind nothing but the conception of an inert mass, or a certain amount of matter, for the determination of which gravitation gives us the means of measuring and comparing. Therefore we may say: To have "a mass of two tons," but not "a force of two tons."

Pressure is a result of this gravitation, and a mass of two tons will exert a pressure of two tons; in this way we may estimate the effect of a spring, hydraulic press, or other similar contrivance, by saying its pressure (not its power) is equal to two tons, meaning thereby that it has the effect, on the material to be pressed, as if two tons weight were placed upon it; but we have in pressure neither force nor power. These conceptions of the latter require other elements, as we shall soon see.

Force is matter in motion, nothing more, nothing less; the abstract idea of force without matter is a nonentity. All the modern discoveries in science tend to prove this more and more plainly. Without matter, force would have no existence, but it may be hidden in matter as molecular invisible motion in the form of heat, electricity, etc. The steam engine, electro-magnetic engine, etc., are there to prove how this molecular motion, or hidden force, may be changed into visible force or motion of matter. Inversely, the caloric friction machine changes motion into heat; the ordinary and also the Holtz electric machine change motion into electricity. In any case, we are driven to the conclusion that all force proceeds from motion of matter, and is finally resolved into motion of matter, either of masses, or into molecular motion, generating one of the so-called imponderable forces.

Chemistry has proved since the last century that the amount of matter in the universe is a constant invariable quantity, and that we cannot create or destroy a single material atom, but can only change its form solid to liquid, or gaseous, or *vice versâ*. So the modern philosophy of mechanics proves that the amount of force (that is, motion of matter) in the universe is a constant quantity, and that we cannot create or destroy the slightest amount of this force, but can only change it from mass motion to molecular motion, that is, heat, electricity, etc., or *vice versâ*.

The measure of force is thus the product of the mass with the distance through which it moves; and as the unit of measure of ordinary masses is the pound, and of distances, the foot, we have adopted the foot-pound as the standard unit of force, meaning "one pound *lifted against gravitation* one foot," not "one pound moved one foot," as we have seen and heard it stated, which of course gave rise to the most absurd calculations in regard to the immense power obtained to drive a steamship or railroad train.

If one pound weight is raised one foot, one unit of force is expended; if, inversely, we cause one pound to descend one foot, we obtain a unit of force back, and may transform this into other mass motion, or into molecular motion. We may cause this mass of one pound to be raised slowly if we have little power to apply, or rapidly if we have greater power; and, inversely, we may cause it to descend slowly, as is done in the weight of a clock, and spend itself gradually during a long period of time, producing slight effects throughout that time; or we may cause it to descend quickly, as is the case with the blow of a hammer, and spend itself during a very short period of time, almost instantaneous, producing a powerful effect for that short time. So the driving in of a nail, which often the pressure of a ton weight would not accomplish, the blow of a hammer of one pound, lasting a small fraction of a second, will accomplish easily. This remark points out forcibly the difference between the weight of masses at rest and of masses in motion; in other words, the immense difference between mere pressure and force.

We may understand from the *Scientific American* that modern discoverers have undiscovered the discoveries of Sir Isaac Newton.

The law of *force* of universal attraction, which was established by Newton, is as follows: *The force of attraction between any two masses is proportionate to the product of the masses, and inversely as the square of their distance apart.*

There is no motion in this definition of force, and it makes no difference whether either one or both the masses are at rest or in motion, the *force* will always hold good with the definition. Call $F =$ force of attraction, between the masses M and m, and $D =$ their distance apart.

$$\text{Then } F = \frac{Mm}{D^2} \quad . \quad . \quad . \quad . \quad . \quad 11.$$

$$FD^2 = Mm \quad . \quad . \quad . \quad . \quad . \quad 12.$$

$$F : M = m : D^2 \quad . \quad . \quad . \quad . \quad . \quad 13.$$

Call $V =$ the mean velocity in the time T, in which the masses M and m would be drawn together by their own attraction, and we have $D = VT$ and $D^2 = V^2T^2$.

$$\text{The work } K = FVT = \frac{Mm}{VT} \quad . \quad . \quad 14.$$

The object of this treatise does not involve the development of this very interesting subject, which must be deferred to a more appropriate occasion.

In the application of the physical elements to practice, we must assume units of measures for each of them, namely:

$F =$ force in pounds, acting on the mass M.

$V =$ velocity in feet per second.

$T =$ time in seconds, in which the force F acts on the mass M, and produces the velocity V.

$W =$ weight in pounds of a body, or of the mass M.

When the body falls freely under the action of gravity, then $F = W$.

$M =$ mass, which is proportionate to weight when compared in one or the same locality.

The mass of a body is said to be its weight divided by the acceleratrix g, or 32.17, for which the unit of mass should be a quantity of matter weighing 32.17 pounds.

There is no name adopted for this unit of mass, the want of which makes the subject obscure. If we are told that the weight of a body is 20, we naturally ask what twenty? and we cannot conceive the magnitude of the weight until we know the name and magnitude of its unit. Although we may know that thirty-two pounds of matter is a unit of mass, it does not make the clear impression as if the unit had a name, for which reason I would propose to christen this unit with the name *mat.*, from the word matter.

Then, the unit *mat.* means a quantity of matter weighing 32.17 pounds, and there will be 69.63 mats in a ton weight of 2240 pounds. Now, we can apply our reasonings to practice.

The actual force F in pounds of attraction between any two masses M and m, expressed in mats, and $D =$ their distance apart in feet, will be closely approximated by the formula—

$$F = \frac{M\,m}{5274460 D^2}. \qquad \qquad 15.$$

ILLUSTRATION OF DYNAMICS OF MATTER.

Let a force F be applied on a mass M at rest, but free to move; the mass will then be set in motion with an accelerated velocity as long as the force acts. If the force F is constant, the acceleration of velocity will also be constant, and the velocity V, attained in the time T, will be from the fundamental analogy of dynamics of matter 7.

$$V = \frac{FT}{M} \quad . \quad . \quad . \quad . \quad 16.$$

That is to say, the force F acting upon the mass M, in the time T, will generate the velocity V.

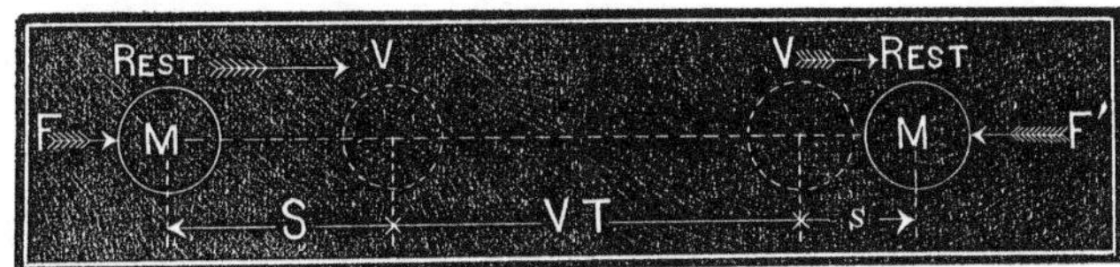

Example.—A force $F = 8$ pounds, is acting upon the mass $M = 4$ mats, for a time of $T = 3$ seconds. Required the velocity V.

$$V = \frac{8 \times 3}{4} = 6 \text{ feet per second.}$$

In the functions 4, 5, and 6, V means the mean velocity in the time T or space S, and when V means the actual velocity of a moving body accelerated from rest to V, or retarded from V to rest, the mean velocity will be ½ V. Then we have the real work expended on the moving mass, or generated in bringing the mass to rest, to be,

$$K = \frac{MV^2}{2} = \frac{FVT}{2} \quad . \quad . \quad . \quad . \quad 17.$$

The space $S = \frac{VT}{2}$, which inserted in formula 17 will be the

$$\text{Work } K = \frac{MV^2}{2} = FS \quad . \quad . \quad . \quad 18.$$

Referring again to the illustration, the force F acted on the mass M only in the space S, after which the body will continue with uniform velocity and generate the space VT, when another force F', independent of F, is applied in opposite direction to stop the motion of the mass M, which is accomplished in the space s, independent of the first space S, but the work consumed or generated will be alike in both cases, namely,

$$K = \frac{MV^2}{2} = FS = F's \quad . \quad . \quad . \quad 19.$$

$$\text{Of which } F : F' = s : S \quad . \quad . \quad . \quad 20.$$

In the case of a falling body, the force F is equal to the weight of the body, and when it strikes a blow, the mean force of resistance is F', acting in the space s, but as the force F or weight of the body is constant even through the space s, the falling body is acted upon by two opposite forces F and F', for which the space S of the fall must also include the space s of the blow.

Then we have the force of a falling body equal to its weight multiplied by the whole height of fall, and the product divided by the space of the blow.

When the force of resistance is irregular, the mean force will still be as the above rule.

In the case of the problem of driving a nail into a piece of wood, as before illustrated, whether accomplished by the aid of a lever or by the weight falling the same space directly upon the nail, the rule will hold good in both cases.

In the case of a steam hammer weighing $W = 3$ tons, and falling $S = 4$ feet, including the compression of the iron by the blow, say, $s = 0.1$ of a foot, required the force of the blow.

$$F' = \frac{WS}{s} = \frac{3 \times 4}{0.1} = 120 \text{ tons.}$$

This is the answer to the original question of Mr. J. E.

The force of resistance F' is not uniform in the space s of compression of the iron forged. It is generally smallest when the hammer first touches the iron, and greatest at the moment the hammer stops, but the mean force throughout the compression s, will be according to the formula.

The *Scientific American* attacked me on the subject of dynamics some seven years ago, as follows:

[*From the Scientific American, June* 29, 1865.]

Work and Power.

In the pages of the *Journal of the Franklin Institute,* a discussion is going on between De Volson Wood, Professor of Civil Engineering in the University of Michigan, and J. W. Nystrom, Acting Chief Engineer, U. S. N., on the subject of work, force and power. The main purpose of Mr. Nystrom seems to be to deny the position that work is independent of time, and he succeeds in involving the question in considerable confusion. The facts of the case are simple and plain enough.

Work is the overcoming of mechanical resistance of any kind, either by raising a weight, dragging a body along, turning off a shaving, or by any other action. The question whether it is independent of the time depends entirely upon the meaning of the language employed. A foot-pound of work is the raising of one pound of matter one foot in vertical height, and this foot-pound is precisely the same quantity whether one second or one thousand years be consumed in the operation.

We may say that a machine is doing the work of raising one foot at the rate of one inch per second; then the work done by the machine will depend upon the time that it is in operation; it will take it twelve seconds to do one foot-pound of work, and twenty-four seconds to do two foot-pounds.

In this case, however, we have attached to the word "work" a meaning for which the word "power" is employed by the standard writers on mechanical philosophy. To keep our ideas clear, it is better to regard the machine as exerting a power of one inch-pound per second, and to confine the word "work" to the aggregate resistance overcome.

One writer argues that 2 and 2 do not always make 4, sometimes making 22. By analogous tricks of language we may confuse our minds in regard to any problem whatever; but a more useful aim of discussion is to free our minds from confusion, and to accomplish this one of the most important steps is to use words always in their exact signification.

Regarding work as the overcoming of physical resistance, it is plain that the aggregate amount of any given quantity is independent of the time required for its performance.

There is probably no higher authority on the philosophy of mechanics than Arthur Morin, and from his "Leçons de Mécanique Pratique," translated by Bennett, we take the following extract:

"*The Idea of Work is Independent of Time.*—We see from what precedes that in the measure of work we have only regarded the effort exerted, and the space described in the direction peculiar to this effort. It is, therefore, independent of time.

"Thus, in raising goods the effect is not measured by the duration of labor, but by the product of the load into the height of its elevation."

The editor of the *Scientific American* declined at first to publish my reply to the above criticism, and he wrote me a letter advising me to consult some authority on the subject, indicating that my views on dynamics were all visionary; whereupon I called at the office of the editor and insisted upon it that my reply should be published, which was at last consented to, and which is as follows:

[*From the Scientific American, September* 9, 1865.]

Force, Power, and Work.

(For the Scientific American.)

Force is a mutual tendency of bodies to attract or repel each other. Its physical constitution is not yet known. We only know its action, which is recognized as pressure and measured by weight. The unit of weight being assumed from the attraction of the earth upon a determined volume of any specific substance; for example, the force of attraction between the earth and 27.7 cubic inches of distilled water, at the temperature of 39.8° Fahr., in an atmosphere balancing 30 inches of mercury, at the level of the sea, which is called one pound avoirdupois. Force is the first element of Power and Work, and can be likened to length, which is a primary element in geometry. Force will here be denoted by the letter F, expressed in pounds.

Velocity is the second element of Power and Work, and may be likened to breadth in geometry. It is that continuous change of position recognized as motion, and is here denoted by the letter V, expressed in feet per second. Velocity is a simple element, although it appears to be dependent on time and space, but the space is divided by the time, and therefore both relieved from the velocity.

Time is the third element of work, and may be likened to thickness in geometry. It implies a continuous action recognized as duration. Time is here denoted by the letter T, expressed in seconds.

POWER is a function of the two first elements—force F, and velocity V—as area in geometry is a function of length and breadth. Power is here denoted by $P = FV$, which means that the power P, is the product of the force F, multiplied by the velocity V. The power so obtained is expressed in foot-pounds, and called dynamic effect, of which there are 550 in a horse-power; or if the velocity is measured in feet per minute, there will be 33,000 foot-pounds in a horse-power. Power is independent of space and time, but it has often been confounded with work, which essentially depends on time and space.

SPACE is a function of the second and third elements—velocity V, and time T—and may be likened to a cross section of a solid, which is a function of breadth and thickness. Space is here denoted by $S = VT$, which means that the space S, is the product of the velocity V, and the time T, expressed in linear feet.

WORK is a function of the three elements,—force F, velocity V, and time T. It may be likened to a solid in geometry, which has the three dimensions,—length, breadth, and thickness. Work is here denoted by $K = FVT$, which means that the work K, is the product obtained by multiplying together the three elements—force F, velocity V, and time T.

Work may also be denoted by $K = FS$, or the product of the force F multiplied by the space S, where it appears as if the work was independent of time, but the time is included in the space $S = VT$.

Work may also be denoted by $K = PT$, which means the power P, multiplied by the time T. Either of the three cases expresses the work in foot-pounds.

Force, velocity, and time are simple physical elements.

Power, space, and work are functions or products of those elements.

The *Scientific American* is read by most mechanics in this country, and it may be further said that that journal is met with in most parts of the world. It evinces a habitual and sincere desire to furnish its readers with correct and instructive articles on scientific subjects, in consideration of which it would be a neglect of duty on my part to pass over in silence its articles on "Work and Power," published on page 71 of the present volume. In that article you say, "The main purpose of Mr. Nystrom seems to be to deny the position that work is independent of time, and he has succeeded in involving the question in considerable confusion." And you think "the facts of the case are simple and plain enough." You proceed to give an antithetical description of what work is, and say, "In this case, however, we have attached to the word *work* a meaning for which the word *power* is employed by the standard writers on mechanical philosophy." Now you will allow me to remark that, in this expression, you have, together with the standard writers, confounded *work* with *power*. You have thus not followed your own good advice, namely, "to free our minds from confusion" by taking "most important steps to use words always in their exact signification." You then go on to say, "Regarding work as the overcoming of physical resistance, it is plain that the aggregate amount of any given quantity is independent of the time required for its performance." Do you not here convey the idea that *work is independent of what it requires*, namely, the *time?* You evidently mean to say that a given quantity of work may be performed in any desired length of time; but you do not seem to conceive that the work is dependent on whatever time is required for its completion.

Referring to a geometrical figure, you may be able to comprehend the position of your argument about work, which is substantially this—*the cubic content of a pancake is independent of the thickness required to make it up!* You say, "The question whether it (work) is independent of the time depends entirely upon the meaning of the language employed." To this I respectfully object, inasmuch as I recognize only one meaning in the language I have employed; and, indeed, the entire controversy upon this subject appears to have sprung from a rejection or misappreciation of the specific meaning I have struggled to establish to the terms *force*, *power*, and *work*. But supposing, for the sake of argument, that my general language is not sufficiently clear, if you can read my algebraical formulas you will not misunderstand me. Why, therefore, do you not exemplify your argument upon my formulas and thereby show its effect in practice? To say that "work is independent of time," is to say that work is dependent of no time, or that any amount of work may be performed in no time—a proposition which is not yet realized.

Work, as before stated, is the product of the three elements, —force, velocity, and time. For a given quantity of work, either one or two of these elements can vary *ad libitum*, but only at the expense of the remaining two or one. Work is thus not confined to any specific relation or ratio to either of those elements, but independent of either one of them it ceases to be work.

I am well aware that the standard authors have to this day considered work independent of time, and they have also confounded force, power, and work with each other, so that we are yet thrown upon our individual authority to decide which is right. Thus, when you and Professor Wood cannot defend your position, it may be very convenient to assert that I use "cant phrases" when I exemplify your arguments, and that "by analogous tricks of language we may confuse our minds in regard to any problem whatever." I nevertheless trust that I have given both perspicuity and precision to my language and meaning in this article, and sincerely hope that it may be tributary to "the consummation—devoutly to be wished for"—of reducing to certainty and system the future reasonings of the scientific world on this subject.

JOHN W. NYSTROM.

[*From the Scientific American, September* 9, 1865.]

Nystrom on Work and Power.

We have a kindly feeling towards Mr. Nystrom, having received from him several valuable contributions. It has seemed to us, however, that his method of explaining the difference between work and power was calculated rather to confuse than to elucidate the subject. In reply, he forwards us a communication containing his explanation, with a request that we would lay it before our readers and let them judge for themselves. We comply with his request with pleasure, and the communication will be found on another page.

The raising of one pound of matter one foot in vertical height is one foot-pound of "work." The raising of 33,000 pounds one foot is 33,000 foot-pounds of work, whether one minute or one hundred years be consumed in the operation.

The power—either of a steam-engine, waterfall, or animal—that can raise 33,000 pounds one foot in each minute of time is one-horse power; the power that can raise 33,000 pounds in half a minute is two-horse power; and the power that can raise 33,000 pounds in one-tenth of each minute is ten-horse power.

Morin and other writers, therefore, say that the idea of work is independent of time, but that time is an element in the measure of power. It seems to us that these writers are correct. It seems to us, also, that the matter is extremely plain and simple.

Mr. Nystrom, on the other hand, while accepting, if we understand him, the above illustrations of both work and power, denies that work is independent of time, or that time is an element of power, and asserts that the subject is not generally understood even by educated engineers. We have criticised his arguments upon it as calculated rather to confuse than elucidate it. From this criticism he wishes to appeal to the judgment of our readers,—an appeal in which we cheerfully concur.

I do not doubt the kindly feeling of the editors of the *Scientific American* towards me, which has been proven in their very kind advice; and I am convinced that they have no improper motives in attacking me on dynamics; but the misfortune is that they do not understand the subject.

In regard to work being dependent or independent of time, we have only to refer to the operation of a steam-engine of a definite power. Whatever time the engine is running, say one hour or one year, the power is constant, independent of time;

but the work executed by that engine depends entirely upon the time of operation.

The best joke of all is that the editors of the *Scientific American* are not ashamed of themselves for undertaking to give me advice on the subject. I have lately received a letter, signed Munn & Co., praising Dr. Vander Weyde and advising me to drop the subject, evidently with the intention of frightening me.

The publication of this pamphlet is the result of their last advice. I have heretofore never entertained the slightest idea of giving advice to the editors of the *Scientific American*, but now I feel disposed to do so, namely:

I will hereby advise the editors of the *Scientific American* to inform themselves well on the subject of dynamics, before they give me advice and attack me on that subject.

I am informed on good authority that Dr. Vander Weyde is engaged for the *Scientific American* to write on scientific subjects or as an acting editor, and he writes articles which are published under the head of correspondence, stating that "the editors are not responsible for the opinions expressed by their correspondents."

It will be noticed in this pamphlet that the *Scientific American*, as well as myself, is attempting to clear up the subject of dynamics in regard to definite meanings of its terms, and we are of different opinions about the true meaning of the term force. In the year 1865 I submitted a communication on the subject of dynamical terms to the National Academy of Sciences, which met at Washington that year; but the papers were handed back to me immediately, with information that "the subject is all clear."

In conclusion, I most respectfully invite the attention of scientific institutions to my classifications of dynamical terms, that a decision may be given, whether it be adopted or rejected. A prompt action on this subject, in aid of establishing precision to the meaning of dynamical terms, would render a great service to the public.

[*From the Manufacturer and Builder, December, 1872.*]

On Measuring Forces.

The question, if for the measure of a force the mass m must be multiplied by the simple velocity v, or with the square of the velocity v^2, is a very old one, and divided the mathematicians in the beginning of the former century, for the term of forty years, into two parties; one headed by Descartes, who defended $v \times m$, the other by Leibnitz, who adhered to the principle expressed by $v^2 \times m$. Descartes defended his principle on that of the lever, in which a body M of 4 pounds on the shorter arm is lifted one inch for every four inches that a body m of one pound attached to the longer arm descends 4 inches. Leibnitz reasons thus: if a body M of 4 feet* falls from a height of 1 yard, and a body m of one foot* from a height of 4 yards, they acquire at the end of these falls such a velocity as would be sufficient to throw them up to the height from which they started. While now the acquired velocities, according to the laws of falling bodies, are in ratio as the square roots of the heights, the smaller mass m has only acquired double the velocity of the larger mass M, and as this is sufficient to throw it up again to four times the height of the larger mass, it is clear that the force of the small mass is to that of the larger as $v^2m : V^2M$.

The adherents of Descartes were Mairan, Desaguliers, Maclauren, Heinsius, etc.; those of Leibnitz were the brothers Bernouilli, Wolf, Gravesande, Musschenbroek, etc. The history of this celebrated scientific controversy has been described in many details by Kant, Arnold, and Kaestner.

An interesting experiment was devised in favor of the view of Leibnitz, which had acquired the name of the theory of the *vis viva*, or living force, in contradistinction of the view of Descartes, which was denominated *vis inertia*, or dead force, and admitted to be only correct in peculiar cases. This experiment was to project balls of equal size into a soft substance, such as wet clay, when it was invariably found that the depth of the hole was proportionate, not to the velocity, but to the square of the velocity of the penetrating body. The same was proved to be the case with a hammer driving in nails; a hammer of 4 pounds moved with a velocity at the rate of 1 foot in a quarter of a second, did produce no more effect than a hammer

* *Long*, I suppose (N).

of 1 pound moved with double the velocity. In short, all the practical experiments proved the correctness of Leibnitz's views, notwithstanding that it could not be denied that the number of units of foot-pounds obtained by multiplying the weight of a mass, or the equivalent pressure of springs (steam in our days), with the single velocity was also the correct expression for measuring forces.

Finally, the two views were reconciled toward the end of the last century, and it was proved that both were right, or that $v + m$ and $v^2 + m$ were both correct, according to circumstances and the conditions assumed.

It was pointed out that the adherents of the theory of Descartes made the silent condition that the effects of the forces were supposed to be obtained in equal times, and did not consider the spaces in which these effects were spent; while the adherents of Leibnitz inversely did not consider the difference of the times in which the effects were obtained, and only considered the spaces, showing the effect produced.

For, not to fatigue the reader with a long string of algebraic formulæ, deduced from the laws of falling bodies, we will only give the results arrived at by J. R. Schmidt, Professor of Mathematics in the Military Academy in Holland, who in *Text-Book of Dynamics* published 1825, has perhaps exposed the matter mathematically in the clearest light; calling the forces of two bodies b and B, the velocities v and V, the masses m and M, the spaces s and S, and the times t and T, he obtains the proportions

$$b : B \div \frac{v\ m}{t} : \frac{V\ M}{T} \div \frac{s\ m}{t^2} : \frac{S\ M}{T^2} \div \frac{v^2 m}{s} \div \frac{V^2 M}{S}$$

making $t = T$, we have

$$b : B \div v\,m : V\,M;$$

that is, in words: *the moving forces are in ratio as the products of the masses with the velocities*, namely, on the conditions that the times in which the effects are manifest are equal.

If we make $s = S$, we have

$$b : B \div v^2\,m : V^2\,M;$$

that is: *the moving forces are in ratio as the products of the masses with the squares of the velocities.* That is to say, on the condition that we understand by velocities those which the bodies have attained after having passed over equal spaces during the continued operation of the forces acting on them, regardless of the time.

He closes the chapter with the following sentence: "The first of these expressions is called the unit of force of Descartes, the latter that of Leibnitz. It is evident that both are equally mathematically correct, but are founded on different premises, and must be used *according to circumstances.*

Note to the above.—The preceding article has been prepared for the special information of those who, like a certain civil engineer of Philadelphia, think that the expressions v + m and $V^2 + M$ can not both be true, finding fault with the statement made by us elsewhere, and proved in the above article, that both expressions *are correct according to circumstances.* Said engineer, in a recent publication, in place of pointing out the errors of his adversaries, descends into personal abuse, and at the same time makes a confession of ignorance in the history of the subject, about which he assumes authority, by indulging in the following joke: "I would ask Dr. Vander Weyde if the circumstances depend upon the weather?"

Remarks.

The above article throws no light "on measuring forces," but only proves that the subject was not understood by either of the parties, DESCARTES or LEIBNITZ.

The article also proves that Dr. Vander Weyde has no clear conception of his own on that subject, and that he merely depends upon the authority of his countrymen. He does not even understand the use of algebraic signs, but uses + for ×, ÷ for =, and ÷ for :.

If $b : B = m\,v : M\,V$, and $b : B = m\,v^2 : M\,V^2$

then $m\,v : M\,V = m\,v^2 : M\,V^2$,

which reduces itself to $v = v^2$ or $V = V^2$,

which cannot be admitted.

In the one case, B means *Momentum;* and in the other case, B means *Work*, which are two distinct functions, neither of which is Force under any circumstance.

It is erroneous and confusing to denote two distinct functions by the same letter B.

The functions $M\,V$ and $M\,V^2$, are so inseparably associated together that one of them cannot exist without the other; and to say that either one of them can be force, according to circumstances, is simply absurd; because no circumstance whatever can have the slightest effect upon their nature.

The momentum $M V$, divided by time, will be force; but that does not prove that $M V$ is force.

The so-called *Vis-viva* $M V^2$, divided by space, will also be force, but does not prove that $M V^2$ is force.

The area of a rectangle divided by either one of its sides, gives the other side, but does not prove that that other side is the area.

Assuming $t = T$ and $s = S$ according to circumstances, is a mathematical trick by which any hypothesis can be sustained. For instance, assume any two quantities, $V W$ and $Q D$ which are to be proved to be alike; we can set up the formula

$$x(V W - Q D) = x(V W - Q D),$$

which cannot be denied.

$$\text{Then } x V W - x Q D = x V W - x Q D,$$

$$\text{and } x V W - x V W = x Q D - x Q D.$$

$$\text{or } V W (x - x) = Q D (x - x).$$

Eliminate the factor $(x - x)$, and the result will be

$$V W = Q D,$$

which was to be proved.

Dr. Vander Weyde's writings on dynamics serve, however, as records to illustrate the chronic confusion which besets this subject.

[*From the Scientific American, Nov. 23, 1872.*]

Intolerance in Science.

We have received a pamphlet entitled "On Force of Falling Bodies and Dynamics of Matter, classified with precision to the meaning of dynamical terms, by John W. Nystrom, C. E." It contains 29 pages, of which, to our disappointment, we find 20 filled with different articles published in 1865 and 1872 in the SCIENTIFIC AMERICAN, only 5 pages of explanation of the author's views on the subject, while the remaining 4 are filled, not with scientific refutations, but with personal abuse of his antagonists, who appear to be very numerous, and from among whom he especially singles out Dr. Vander Weyde, saying: "It will do no harm to tell the truth to one of them, every now and then . . . . equally applicable to all the rest of the high authorities who have invariably attacked me. . . When my ideas differ from what is written in their books, they blindly suppose that I am wrong," etc. He further threatens that he will warn the university where Dr. Vander Weyde graduated of his erroneous philosophy, and "if that university cannot sustain its doctor's statements, he ought to be called back and made to study over again, or be requested to return his doctor's diploma."

We have already, in our paper of July 29 and September 9, 1865, concerning Mr. Nystrom's views, given our opinions; they agree perfectly with those of the National Academy of Sciences, which met in Washington that same year, and would not accept Mr. Nystrom's papers on that subject, as his method of explanation rather confused than elucidated the matter in question; we are, therefore, not inclined to go into any argument at present, but will only remark that it strikes us as not a little curious that Mr. Nystrom finds so much fault with Dr. Vander Weyde's disagreeing with the books and accepted views, while Mr. Nystrom himself boastfully proclaims that the books and accepted views are erroneous; thus he is guilty of the same offence. Only the manner differs in which both gentlemen disagree from the books, and this appears to be very distasteful to Mr. Nystrom.

We are aware that in theological colleges the diplomas are sometimes withdrawn when the graduates preach heresies, not sanctioned by their orthodox *Mater Alma;* but we wish to remind Mr. Nystrom that science is eminently tolerant, and that a graduate, after having been taught the prevalent scientific doctrines in college (and we are convinced that this was the case with Dr. Vander Weyde) is at full liberty to promulgate afterwards new scientific ideas or philosophies, without fear of being prosecuted, called back, or having his diploma annulled. On the contrary, such attempts are considered praiseworthy, as without them science would not progress; we are, therefore, far from blaming Mr. Nystrom for trying to promulgate and defend his views, only he must acknowledge that others have a right to the same privileges, which nobody wishes to deprive him of, even if they cannot agree with his peculiar notions, whether they be on velocity of thunder (see SCIENTIFIC AMERICAN of August 24, 1872), the decimal and tonal systems, or the force of falling bodies, etc.

Answer to the Above.

I am only acting on the defensive, and since the *Scientific American* assures me that, "in Dr. Vander Weyde, I have a foeman worthy of my lance," (page 14,) I have no apology to spare.

My assailants have attacked me from all directions on the subject of dynamics, for the last ten years, in which time I have treated them with great consideration, until my patience became exhausted, and inasmuch as they are very strong, both in numbers and position, whereby my elements of dynamics have been withheld from the public, it would seem both justifiable and necessary to depart a little perhaps, from the *Suaivter in modo, fortiter in re.*

One of my objects has been to convince my opponents that I do not take their statements for granted, merely because they come fortified by high authority, for it is my custom to analyze the *substance* of an argument, without regard to the source from which it comes, and have thus disappointed my opponents who are accustomed, it would seem, to regard an official title as a re-inforcement of a fallacy.

I have experienced no difficulty in finding out the *weakness* of an opponent, which is generally exposed by the nature of his argument, and sometimes it has been quite amusing to see how he contorts his meshes to find a mare's nest.

DISCUSSION ON THE SUBJECT OF

DYNAMICS,

AND

ON THE CONFUSION OF DYNAMICAL TERMS,

READ AT THE MEETING OF THE

FRANKLIN INSTITUTE,

September, 1873,

BY

JOHN W. NYSTROM, C. E.

PHILADELPHIA:
JOHN P. MURPHY, PRINTER, FIFTH AND WALNUT.
1873.

DISCUSSION ON DYNAMICS.

MR. PRESIDENT,

AND MEMBERS OF THE INSTITUTE.

My object in reading the following discussion on dynamics at this meeting, is to impress upon the members of the Institute, the importance of appointing a Committee for the purpose of establishing precision to the meaning of dynamical terms.

Discussions on this subject are published in the Journal of the Franklin Institute for the years 1864 and 1865, and which have been revived by a question on "Force of falling bodies," published in the Scientific American, June 8th, 1872. A pamphlet was written on the same subject and published by J. B. Lippincott & Co., Philadelphia. This pamphlet has been distributed amongst all the Colleges and other Scientific Institutions in the United States and some in Europe, so that it is well known. In the year 1865, it was proposed to publish a work on "*Elements of Mechanics, classified with precision to the meaning of dynamical terms*," but the publishers naturally hesitated to publish such a work without a favorable opinion of high Authorities. Upon consultation they were made to believe that those ideas of dynamics were not right.

Repeated efforts have since been made to convince the publishers of the importance and necessity of such a work, but thus far in vain.

In the year 1872, the publishers again consulted a Professor in a distinguished College, in regard to the ideas of dynamics published in the pamphlet on force of falling bodies, and received on the subject the following answer:

THE PROFESSOR'S OPINIONS.

"I have examined the pamphlet on 'Force of falling bodies,' and find much to condemn, and very little to approve.

The original question out of which the whole dispute arose, probably used the word *force* instead of *work* or *effect.*

The first answer by E. E. W. has no meaning.

The second answer by W. H. P. will be right if the word *work* is substituted for *force* throughout.

To Dr. Vander Weyde's assumption that 'weight is not force,' and that 'force is absolutely nothing but matter in motion,' I cannot assent. It would destroy all the laws of statics.

Mr. Nystrom while starting with a true idea of force, confounds *work* and *power*. The former the actual effect, without regard to time, the latter the capability of a machine to produce a certain effect in a certain time.

It is a common error to call 33.000 foot-lbs. one horse power, whereas it is simply an amount of work, and has no reference to time. If a motor of any kind does this amount of work *in one minute,* it has expended one horse power.

Pages 25 and 30 are the only ones in the whole pamphlet to which I should assent, and in them the whole matter is clearly stated.

There is not the slightest confusion about the use of the terms *force, power* and *work* by standard authors; and a work on Mechanics based upon the ideas of either Vander Weyde or Nystrom, as shown by this pamphlet, would not meet the approval of scientific men.

In his answer to the original question Nystrom (p. 24) assumes a penetration of 0.1 of a foot, and thus makes the force 120 tons. But if the same ram had done a different work, in which the penetration was 0.2 of a foot, he would have called the force 60 tons, if 0.3 of a foot, the force would have been 40 tons.

What he is measuring is the mean resistance, and does vary inversely as the penetration. But in this sense, the force of the ram is any thing I choose to make it from zero to infinity, according to the substance upon which it operates. There is nothing new in the pamphlet except its errors."

In consequence of the above opinions, the publishers again declined to publish the writers "Elements of Mechanics," and even kept the opinions a secret for some nine months, when it came to the writer's knowledge by mere accident, or rather by answers to cross-questions. The publishers having confidence in the opinions received on the subject, did not hesitate in giving semi-frank answer, from which it was evident that similar opinions had been received from various sources.

The case confirms the correctness of the writer's statements on page 16 in the pamphlet on "Force of falling bodies," but his opponents could not catch the spirit of those statements.

The publishers were addressed on the subject with the following reply:

To the Publishers.

GENTLEMEN,

Accept my most sincere thanks for the copy of the opinions on my pamphlet on "Force of falling bodies," written some nine months ago by a Professor whose name must be withheld for reasons well understood. I cannot overlook those opinions in silence, and with your permission, will proceed to make a few remarks upon the same.

The Professor's objections are repetitions of what have been made over and over again by Professors of College ideas of dynamics, and which have been answered and defeated in the Journal of the Franklin Institute, and in Scientific American some nine years ago, and they are also answered and defeated in the pamphlet in question. I shall avail myself of our Professor's opinions as basis for discussing the confused condition of dynamics, and hope to be distinctly understood, that my sharp reply is not intended to bear only upon our Professor, who is not responsible for the doctrine laid down before him to teach.

I understand very well what the Professor means, but I am making remarks only upon what he says, because it is his statements which have caused the publication of my book to be postponed.

His opinions will be numbered in the order the Professor gives them, to make the subject clear.

1 { The Professor says, "I have examined the pamphlet on "Force of falling bodies," and find much to condemn and very little to approve."

Unfortunately, the Professor has condemned what is right and approved what is wrong, as will be explained hereafter.

2 { "The original question out of which the whole dispute arose, probably used the word *force* instead of *work* or *effect.*"

I believe that the inquirer meant *force* as he said, but whatever he meant, we must answer the question as it is asked. When the question is force, the answer must be force, otherwise the question is not answered. What reason has the Professor for supposing that the question should have been *work*? perhaps to accommodate the wrong answers.

When Colleges and text-books of standard authors confound *force*, *power* and *work*, we may expect to find questions and answers confounded, as was the case with those on force of falling bodies.

3 { "The first answer of E. E. W. has no meaning."

We often find long articles on dynamics, which have no meaning, but which abound in curious terms, whose signification it may be doubted whether the author himself understands.

4 "The second answer by W. H. P. would be right if the word *work* is substituted for *force* throughout."

Then it would not answer the question, which is *force*, and not *work*.

When the term *force* is used, the answer is not right; and when the term *work* is used, the answer will be wrong.

If the Professor thinks that simply a term would make the answer right, it would be well to select still another term. There is such an abundance of mysterious terms in College dynamics, that he ought to be able to find one that would make the answer right.

Vis Viva is a favorite term among the Professors. **Quantity of motion** or **Quantity of moving force,** might answer the purpose, but even if these terms should fail, his resources are by no means exhausted, and as plain work is defective, **quantity of work** or **total quantity of work,** might answer, but still stronger if necessary, namely, the long term **actual total quantity of work,** may accomplish the object; and he has yet a dozen different terms in reserve, in the form of **mechanical power, mode of motion, mode of force, moment of activity, dynamic effect, quantity of action, mechanical effect, accumulated work, force of motion, living force,** and different kinds of **energy**.

The above confusion of terms has not found, and never will find its way into the machine shop, but is used only in text-books and in Colleges.

I have often observed that when a Professor gets puzzled in dynamics, he employs curious terms like those above, which have no meaning.

5 "Dr. Vander Weyde's assumption that weight is not force, and that force is absolutely nothing but matter in motion, I cannot assent to. It would destroy all the laws of statics."

The writings referred to by the Professor in the last three opinions, are productions of Professors who have evidently had a College education and are familiar with the text-books of the standard authors, and when our Professor intimates that the writings of those men are wrong, if he is correct, the Colleges and text-books must themselves be wrong or incapable of correctly teaching the subject of dynamics, which is actually the case, and of which we frequently meet examples.

6 " Mr. Nystrom while starting with a true idea of force," * *

I stated distinctly in my pamphlet that "I do not acknowledge the existence of more than one kind of force in physics," whilst the Colleges and text-book teach about a dozen different kinds of forces. If my idea of force is true, then the Colleges and text-books must be wrong.

7 " Confounds *power* and *work*." * * * * * * *

The difference between *power* and *work* was discussed in the Journal of the Franklin Institute in the years 1864 and 1865, by Professor De Volson Wood, who although at first enunciated different views, finally acknowledged that I was right. Our Professor if at variance with the writer, is therefore at variance with Professor Wood, in regard to the distinction between power and work.

If the writer is in error in his conception of power and work, then the formulas and examples on that subject in his Pocket Book must be wrong; but they have been successfully in use by engineers for some twenty years, and they will hold good for ever, in spite of the Professor's opinions.

8 " The former (work) is the actual effect,"

Here he gives work another name without defining its constituent elements, which are evidently not clear to him, because he adds,

9 without regard to time,

Then any amount of work can be accomplished in no time, a proposition which cannot be admitted.

10 " The latter (power) is the capability of a machine to produce a certain effect in a certain time."

There is no capability in a machine to produce work or effect, for if there was, the great problem of perpetual motion would be solved.

Work, is produced only by generation or absorption of heat.

A machine only conveys work from one locality to another.

A street-car is drawn by two horse-power,—where is the machine and what is the time?—Does the Professor call a horse a machine?

The work is produced by combustion of the food the horse has eaten, and that combustion produces heat which is work. This work is conveyed by the frame-work of the horse and traces to the car, by which motion is produced.

Work, is the product of *force*, *velocity* and *time*, but if we take the time out from the work, the remainder will be *power*, which is the product of force and velocity. Thus it is perceived that work is dependent upon time, and power is independent of time.

When the Professor says, "in a certain time," which is equivalent to the expression "per unit of time," he takes the time out from the work, but he evidently thinks that he puts the time into the work in order to convert it into power.

I should like to know the difference between the three terms used by the Professor, namely *effect*, *actual effect* and a *certain effect*?

If a certain effect means work, then the work done in a certain time is not work but power; but as no work can be accomplished without a certain time, the term work has no meaning, because it is all power according to the Professor's ideas of dynamics.

Most of the Professors with whom I have conversed on the subject of dynamics, have the same erroneous ideas, namely, that work is independent of time, and that power is dependent upon time, or that power is the work done in a unit of time, which is the doctrine of our present text-books.

If power was the work done in a unit of time, then power would be a portion of work, which is a great error. Power and work are two distinct principles.

11 { "It is a common error to call 33.000 foot-pounds one horse-power," *

There are three kinds of foot-pounds in dynamics, namely, foot-lbs. of work, foot-lbs. of power, and foot-lbs. of momentum.

1st. Foot-lbs. of work is force in pounds multiplied by space in feet.

2nd. Foot lbs. of power, is force in pounds multiplied by velocity in feet per unit of time.

3rd. Foot-lbs. of dynamic momentum is mass expressed in pounds, multiplied by velocity in feet per unit of time.

When the velocity is expressed in feet per minute, there will be 33,000 foot-lbs. of power per horse-power, but when expressed in feet per second, there will be only 550 foot-lbs. per horse-power.

It is a fatal imperfection in text-books on dynamics, not to clearly distinguish these three kinds of foot-pounds.

In the writer's treatise on "Elements of Mechanics," now nine years in manuscript, those functions are clearly distinguished from one another.

12 { "Whereas it is simply an amount of work, and has no reference to time."

No amount of work can be accomplished without time.

A street-car is drawn by two horse-power, without regard to time, but the work accomplished in drawing the car from one station to another, requires time. The longer we toil, the more work will be done, but if we have no time to do the work, it will remain undone.

13 { "If a motor of any kind does this amount of work in one minute, it has expended one horse-power."

Then, if the motor continue to work another minute, it must have expended another horse-power, or it will expend one horse-power per minute, or 60 horse-power per hour, according to the Professor's explanation on the subject.

How much horse-power has the motor expended when half that work, or 16,500 foot-pounds is accomplished in half a minute? The correct answer is *one horse-power.*

How much horse-power has the motor expended when a work of 550 foot-pounds is accomplished in one second? The answer is *one horse-power.*

We see here that the horse-power is independent of the time of action, and that the work is proportionate to the time. It is the power which does the work, and not the motor.

The power multiplied by the time of action, gives the work done.

The work divided by the time in which it is executed, gives the operating power in doing that work.

If the Professor had dissolved the functions of *power* and *work* into their constituent elements, we could have formed a clear conception of the difference between the two functions; but he confuses the elements, and proves that he does not understand their combination.

The Colleges and text-books make no distinction between elements and functions in dynamics, for which reason the subject is confused and difficult to learn and comprehend, as chemistry would be, if there was no distinction between material elements and their binary compounds. The Professor's distinction between power and work is, that *power* is capability, and *work* is effect.

14 { "Pages 25 and 30 are the only ones in the whole pamphlet to which I should assent, and in them the whole matter is clearly stated."

In that case, the writer's explanation of "Force of falling bodies," which is not in the pages 25 and 30, must be wrong; and that problem remains yet to be solved.

The Professor did not solve the problem, and it is not solved in any text book.

There is nothing said about Force of falling bodies or dynamics of matter, in the pages referred to; still the Professor says that "in them the whole matter is clearly stated."

The pages 25 and 30 are written by the Editors of the Scientific American, and contain the stereotyped confusion from text-books, and which are disapproved in the same pamphlet.

The Professor has thus condemned what is right and approved what is wrong.

15 "There is not the slightest confusion about the use of the terms *force*, *power* and *work* by standard authors."

Why then do they answer *work* to a question of *force*? They discuss in the writer's pamphlet whether momentum MV or *Vis-viva* MV^2 is force; and they say that both are forces under different circumstances, whilst the fact is that neither of these functions is force under any circumstance, but the Colleges and text-books teach that they are forces. The Professor himself proposes to use the term *work* in answer to a question of *force*.

On the very same page 25, in the pamphlet which the Professor has approved to be correct, a statement ran thus:

"In this case however, we have attached the word *work* in a meaning for which the word *power* is employed by standard authors," and the same writer proposes to "call power *foot-inches* and work *foot-pounds*!"

Mr. W. H. Pratt, remarks on page 7 in the same pamphlet, that "The general confusion of ideas upon this subject is probably largely due to the fact that the text-books differ widely, and the majority of them are entirely wrong, as they almost all teach that the striking force is proportional to the velocity, whereas it is in fact, proportional to the square of the velocity, as is readily shown by the law of falling bodies, enunciated in the very same book."

We frequently hear and read about confusion of dynamics in our text-books. Blunders in machinery are frequently made at enormous expense, for the very reason that the subject of dynamics cannot be correctly learned from our present text-books.

The London Engineer of January 10th, 17th, 24th and 31st, of this year 1873, contains long editorial articles headed "The use and abuse of dynamical terms."

The Editor points out the confusion of dynamics in a great number of works of standard authors, and says that Professor Faraday was a flagrant offender in carrying the abuse of dynamical terms to its highest degree. Professor Tyndal's expression, "heat as a mode of motion" was originated by Professor Faraday.

The "Engineer" attempts to clear up the subject of dynamics of matter, and makes the following statement:

"Of little scientific use or value is the term *momentum* of a body —an expression very often, and much too often, loosely used for *Vis-viva*, or accumulated work."

The term *momentum* like any other term, is of no use to any one who does not understand the use of it.

The dynamic momentum of a body, means the product of the force and time consumed in bringing the body from rest to motion, or from motion to rest.

Momentum divided by time, gives the force required to stop or set the body in motion in that time.

Momentum divided by force, gives the time in which the body is set in motion or brought to rest.

The term *momentum* is of equal importance to any other term in dynamics.

The Editor of the Engineer says, "It is often very evident that the writer himself not seldom gets entangled in his own loose definitions and language. A remarkable exception to this common rule was formed by Professor Macquorn Rankin, who, sparing no trouble to afford the clearest conception of the meanings of the terms he employed, might also be said to have framed the modern English nomenclature used in the best works on applied mechanics."

Professor Rankin's works might justly be said to be of the highest scientific character, and which would be of great importance if not for their confusion of dynamical terms which renders his works of little utility, because very few can understand how to read them.

Questions involving dynamics of matter, are frequently asked, but rarely ever answered correctly, and when an answer happens to be correct, it is generally made with such curious terms that it requires high scholarship to understand it; and thus, the knowledge of dynamics is limited to but few specialists.

When a correct answer is couched in simple and proper terms, which can be understood by those who have not had a College education, then the Professors do not understand it. Accordingly, when the writer produces an article on dynamics expressed by simple and proper terms, the Professors blindly suppose that he is wrong, and attack him with irrelevant philosophy.

The Editors of the Scientific American have labored to find out the philosophy of the fly-wheel, but have not succeeded for the very reason that the difference between *force*, *power* and *work*, is not well understood,

We often meet with the expression "horse-power per minute," or "horse-power per hour," which has no meaning. Power is independent of time, but it is correct to say "work per hour or minute." The expression "consumption of fuel per horse-power per hour" should be expressed "consumption of fuel per hour per horse-power." It is the fuel which is divided by the time and not the horse-power. The consumption of fuel is work, which divided by time gives power.

16 "And a work on mechanics, based upon the ideas of either Vander Weyde or Nystrom, as shown by this pamphlet, would not meet the approval of scientific men."

The writer's ideas of dynamics are already approved by the few scientific men who understand them, and they will work their own way with engineers and mechanics who have not been confounded with College dynamics, for they understand those ideas at the first glance, whilst the Professors have not been able to comprehend the subject during a consideration of nine years.

Under such circumstances, we ought to publish the book for the benefit of the public, and leave the Professors behind.

The Colleges will at length be obliged to adopt those ideas, for they will not continue to teach a confusion of dynamics, after the subject is arranged into a definite system.

The present conglomeration of dynamics in text-books, is a great and unnecessary burden upon professors to teach, and upon students to learn, and when once acquired, it is soon lost; but when the subject is brought down to its simple form, it is easily acquired, and more firmly fixed in the memory.

Mr. Hugo Bilgram stated in the Scientific American (October 5, 1872), that "Mr. Nystrom uses different words, which can be better understood by the workmen."

A simple and well-known term, which expresses the true meaning, is used for each element und function; while the Colleges and text-books employ a confusion of terms without much distinction to which quantity they are used, which, when brought into practice, we find unavailable and incomprehensible.

17 "In his answer to the original question, Nystrom (page 24) assumes a penetration of 0.1 of a foot, and thus makes the force 120 tons. But if the same ram had done a different work in which the penetration was 0.2 of a foot, he would have called the force 60 tons, if 0.3 of a foot the force would have been 40 tons."

Certainly, the Professor is right,—that is what the quoted formula gives, and such is the nature of the case.

He does not say directly that the formula is wrong, but he puts his statement into such an equivocal shape as to convey the idea that it is not quite right, and that he himself understands the subject better.

18 "What he is measuring is the mean resistance and does vary inversely as the penetration, but in this sense the force of the ram is anything I choose to make it from zero to infinity, according to the substance upon which it operates."

The answer to the 18th opinion is precisely the same as that to the 17th. Those opinions convey no additional meaning whatever, beyond that which is expressed by the formula.

There is no other means of measuring the force of the ram but by the force of resistance. It is the force of the ram which drives the nails into the wood, and not the force of resistance.

19 "There is nothing new in the pamphlet except its errors."

When the Professor read this pamphlet, he evidently discovered many ideas of dynamics which are not to be found in his books, and considering them errors, he consequently condemned them because it would not be safe for him to approve that which is contrary to his faith.

The dynamic ideas on pages 25 and 30, agree with the confusion in the Professor's books, and which he accordingly felt safe to approve.

Professor De Volson Wood, in a remark on an article on dynamics, published in the Journal of the Franklin Institute, says "I observe that Nystrom's reasoning upon all the new matter of the article is remarkably correct, and his ideas more definite." Consequently our Professor does not agree with Professor Wood. Both cannot be right.

If the Professor could be tempted to come forward in an open contest on the subject of dynamics, and defend his views if he can, the interest of truth alone could be promoted to the general edification of all who appreciate the importance of this subject; but I am sure that he dare not so expose himself, and he will likely give an excuse for not accepting such an invitation, from which you can safely conclude that his opinions are worth nothing except in deceiving you. Very Respectfully,

JOHN W. NYSTROM.

Philadelphia, Aug. 23, 1873.

The publishers sent a copy of the above arguments to the Professor mentioned, with an invitation to discuss the subject openly. After a week's consideration, the Professor naturally declined the challenge, but he still insisted that his opinions were correct, and that his cognito should be strictly maintained, whereupon the publication in question, of the " Elements of Mechanics," was indefinitely postponed. Men who are considered as high authorities on the subject, thus deceive the publishers with false opinions.

The members of the Institute are well aware of the difficulty in learning dynamics from our present text-books, the reason of which is exposed in the above arguments.

In hope that the Franklin Institute will take an active part in the elucidation of this important subject, the following motion is respectfully submitted for the members to act upon.

Motion by Mr. Nystrom.

Whereas, There exist various conflicting opinions in regard to the true meaning of dynamical terms, and that text-books on dynamics differ widely, and employ a variety of terms which do not express the true meaning for which they are intended, and thus render the subject of dynamics difficult to learn and comprehend, and

Whereas, The subject of classification of dynamical terms originated at the Franklin Institute, and should be disposed of by the same Institution, therefore,

Resolved, That a Committee be appointed by the Franklin Institute, for the purpose of establishing precision to the meaning of dynamical terms, and of selecting and adopting such terms as may be found proper, and of rejecting those that may be considered otherwise.

The resolution was seconded by Mr. C. Chabot, and carried.

The President appointed the following Committee on Dynamical Terms.

JOHN W. NYSTROM, Chairman,
FAIRMAN ROGERS, Professor,
J. VAUGHAN MERRICK, Esquire,
LEONARD G. FRANCK, Professor,
† GEORGE F. BARKER, Professor,
JOHN H. TOWNE, Esquire.

www.ingramcontent.com/pod-product-compliance
Lightning Source LLC
LaVergne TN
LVHW011121110826
845150LV00008B/2207

* 9 7 8 1 4 2 5 5 0 3 4 2 0 *